W9-CTV-839

1980
Yearbook of Astronomy

1980 Yearbook of Astronomy

Edited by

PATRICK MOORE

W·W· Norton & Company Inc
NEW YORK

Copyright © 1979 by Sidgwick and Jackson Limited

First American Edition 1980

*All Rights Reserved. No part of this publication may
be reproduced or transmitted in any form or by any
means, electronic or mechanical, including photocopy,
recording or any information storage and retrieval
system, without permission in writing from the
publisher.*

ISBN 0-393-01318-9

Library of Congress Catalog Card No. 62–1706

Printed in Great Britain

Contents

Editor's Foreword · 7

Preface · 9

PART ONE: EVENTS OF 1980

Notes on the Star Charts · 13

Northern Star Charts · 16

Southern Star Charts · 42

The Planets and the Ecliptic · 68

Phases of the Moon, 1980 · 73

The Planets in 1980 · 74

Events of 1980 · 76

Monthly Notes · 77

Eclipses in 1980 · 127

Occultations in 1980 · 129

Comets in 1980 · 131

Meteors in 1980 · 134

Minor Planets in 1980 · 135

Some Events of 1981 · 137

PART TWO: ARTICLE SECTION

Is Life on Earth Unique? *Iosif Shklovsky* · 141

Meteor Streams and Rainfall *D.L. McNaughton* · 144

Observing the Sun *Peter J. Garbett* · 155

Some Aspects of Martian Dust Storms observed during the Viking Mission *Garry E. Hunt* · 166

Some Properties of Open Clusters *Åke Wallenquist* · 178

The T Tauri Stars *Martin Cohen* · 186

The Stripping Galaxy *David A. Allen* · 193

Gamma-Ray Astronomy *A.W. Wolfendale* · 201

Recent Advances in Astronomy *Patrick Moore* · 217

PART THREE: MISCELLANEOUS

Some Interesting Telescopic Variable Stars · 224

Some Interesting Double Stars · 225

Some Interesting Clusters and Nebulæ · 227

Some Recent Books · 228

Our Contributors · 229

Astronomical Societies in Great Britain · 231

Editor's Foreword

This new edition of the *Yearbook* follows the same lines as its predecessors. The southern hemisphere is dealt with as fully as the northern, and we are delighted to welcome some new contributors – from Durham, Professor Arnold Wolfendale; from Sweden, Professor Åke Wallenquist; from Rhodesia, David McNaughton; and from the Soviet Union, Professor Iosif Shklovsky. To these, and of course to our regular contributors (Drs Hunt, Allen, and Cohen) we are deeply indebted. As in all previous *Yearbooks,* the charts and notes of phenomena for each month have been provided by Dr J.G. Porter. We hope that 'the mixture as before' will continue to be acceptable!

PATRICK MOORE

Selsey, May 1979

Preface

New readers will find that all the information in this *Yearbook* is given in diagrammatic or descriptive form; the positions of the planets may easily be found on the specially designed star charts, while the monthly notes describe the movements of the planets and give details of other astronomical phenomena visible in both the northern and southern hemispheres. Two sets of star charts are provided. The **Northern Charts** (pp. 16 to 41) are designed for use in latitude 52 degrees north, but may be used without alteration throughout the British Isles, and (except in the case of eclipses and occultations) in other countries of similar north latitude. The **Southern Charts** (pp. 42 to 67) are drawn for latitude 35 degrees south, and are suitable for use in South Africa, Australia and New Zealand, and other stations in approximately the same south latitude. The reader who needs more detailed information will find *Norton's Star Atlas* (Gall and Inglis) an invaluable guide, while more precise positions of the planets and their satellites, together with predictions of occultations, meteor showers, and periodic comets may be found in the *Handbook* of the British Astronomical Association. A somewhat similar publication is the *Observer's Handbook* of the Royal Astronomical Society of Canada, and readers will also find details of forthcoming events given in the American *Sky and Telescope*. This monthly publication also produces a special occultation supplement giving predictions for the United States and Canada.

Important Note

The times given on the star charts and in the Monthly Notes are generally given as local times, using the 24-hour clock, the day beginning at midnight. The times of a few events (e.g.

9

eclipses) are given in Greenwich Mean Time (G.M.T.), which is related to local time by the formula

Local Mean Time = G.M.T. — west longitude.

In practice, small differences of longitude are ignored, and the observer will use local clock time, which will be the appropriate Standard (or Zone) Time. As the formula indicates, places in west longitude will have a Standard Time slow on G.M.T., while places in east longitude will have Standard Times fast on G.M.T. As examples we have:

Standard Time in

New Zealand	G.M.T.	+	12 hours
Victoria; N.S.W.	G.M.T.	+	10 hours
Western Australia	G.M.T.	+	8 hours
South Africa	G.M.T.	+	2 hours
British Isles	G.M.T.		
Eastern S.T.	G.M.T.	—	5 hours
Central S.T.	G.M.T.	—	6 hours, etc.

If Summer Time is in use, the clocks will have been advanced by one hour, and this hour must be subtracted from the clock time to give Standard Time.

In Great Britain and N. Ireland, Summer Time will be in force in 1980 from 16 March 02 ^{h}until 26 October 02 ^{h}G.M.T.

PART ONE

Events of 1980

MONTHLY CHARTS and
ASTRONOMICAL PHENOMENA

Notes on the Star Charts

The stars, together with the Sun, Moon and planets seem to be set on the surface of the celestial sphere, which appears to rotate about the Earth from east to west. Since it is impossible to represent a curved surface accurately on a plane, any kind of star map is bound to contain some form of distortion. But it is well known that the eye can endure some kinds of distortion better than others, and it is particularly true that the eye is most sensitive to deviations from the vertical and horizontal. For this reason the star charts given in this volume have been designed to give a true representation of vertical and horizontal lines, whatever may be the resulting distortion in the shape of a constellation figure. It will be found that the amount of distortion is, in general, quite small, and is only obvious in the case of large constellations such as Leo and Pegasus, when these appear at the top of the charts, and so are drawn out sideways.

The charts show all stars down to the fourth magnitude, together with a number of fainter stars which are necessary to define the shape of a constellation. There is no standard system for representing the outlines of the constellations, and triangles and other simple figures have been used to give outlines which are easy to follow with the naked eye. The names of the constellations are given, together with the proper names of the brighter stars. The apparent magnitudes of the stars are indicated roughly by using four different sizes of dots, the larger dots representing the bright stars.

The two sets of star charts are similar in design. At each opening there is a group of four charts which give a complete

13

coverage of the sky up to an altitude of 62½ degrees; there are twelve such groups to cover the entire year. In the **Northern Charts** (for 52 degrees north) the upper two charts show the southern sky, south being at the centre and east on the left. The coverage is from 10 degrees north of east (top left) to 10 degrees north of west (top right). The two lower charts show the northern sky from 10 degrees south of west (lower left) to 10 degrees south of east (lower right). There is thus an overlap east and west.

Conversely, in the **Southern Charts** (for 35 degrees south) the upper two charts show the northern sky, with north at the centre and east on the right. The two lower charts show the southern sky, with south at the centre and east on the left. The coverage and overlap is the same on both sets of charts.

Because the sidereal day is shorter than the solar day, the stars appear to rise and set about four minutes earlier each day, and this amounts to two hours in a month. Hence the twelve groups of charts in each set are sufficient to give the appearance of the sky throughout the day at intervals of two hours, or at the same time of night at monthly intervals throughout the year. The actual range of dates and times when the stars on the charts are visible is indicated at the top of each page. Each group is numbered in bold type, and the number to be used for any given month and time is summarised in the following table:

Local Time	18h	20h	22h	0h	2h	4h	6h
January	11	12	1	2	3	4	5
February	12	1	2	3	4	5	6
March	1	2	3	4	5	6	7
April	2	3	4	5	6	7	8
May	3	4	5	6	7	8	9
June	4	5	6	7	8	9	10
July	5	6	7	8	9	10	11
August	6	7	8	9	10	11	12
September	7	8	9	10	11	12	1
October	8	9	10	11	12	1	2
November	9	10	11	12	1	2	3
December	10	11	12	1	2	3	4

The charts are drawn to scale, the horizontal measurements, marked at every 10 degrees, giving the azimuths (or true bearings) measured from the north round through east (90 degrees), south (180 degrees), and west (270 degrees). The vertical measurements, similarly marked, give the altitudes of the stars up to 62½ degrees. Estimates of altitude and azimuth made from these charts will necessarily be mere approximations, since no observer will be exactly at the adopted latitude, or at the stated time, but they will serve for the identification of stars and planets.

The ecliptic is drawn as a broken line on which longitude is marked at every 10 degrees; the positions of the planets are then easily found by reference to the table on page 74. It will be noticed that on the southern charts the ecliptic may reach an altitude in excess of 62½ degrees on star charts 5 to 9. The continuation of the broken line will be found on the charts of overhead stars.

There is a curious illusion that stars at an altitude of 60 degrees or more are actually overhead, and the beginner may often feel that he is leaning over backwards in trying to see them. These overhead stars are given separately on the pages immediately following the main star charts. The entire year is covered at one opening, each of the four maps showing the overhead stars at times which correspond to those of three of the main star charts. The position of the zenith is indicated by a cross, and this cross marks the centre of a circle which is 35 degrees from the zenith; there is thus a small overlap with the main charts.

The broken line leading from the north (on the Northern Charts) or from the south (on the Southern Charts) is numbered to indicate the corresponding main chart. Thus on page 40 the N-S line numbered 6 is to be regarded as an extension of the centre (south) line of chart 6 on pages 26 and 27, and at the top of these pages are printed the dates and times which are appropriate. Similarly, on page 67, the S-N line numbered 10 connects with the north line of the upper charts on page 60.

The overhead stars are plotted as maps on a conical projection, and the scale is rather smaller than that of the main charts.

1L

October 6 at 5^h	October 21 at 4^h
November 6 at 3^h	November 21 at 2^h
December 6 at 1^h	December 21 at midnight
January 6 at 23^h	January 21 at 22^h
February 6 at 21^h	February 21 at 20^h

October 6 at 5^h	October 21 at 4^h
November 6 at 3^h	November 21 at 2^h
December 6 at 1^h	December 21 at midnight
January 6 at 23^h	January 21 at 22^h
February 6 at 21^h	February 21 at 20^h

1R

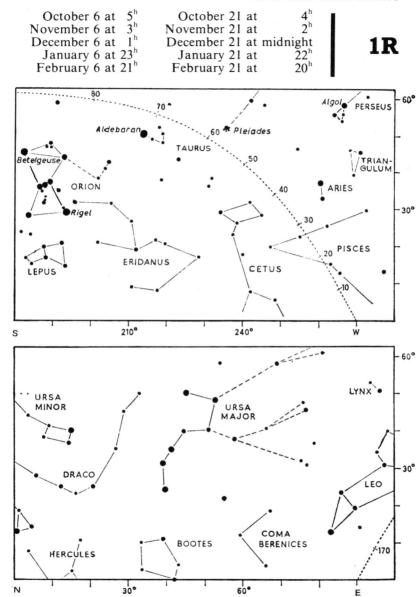

17

2L

November 6 at 5ʰ   November 21 at 4ʰ
December 6 at 3ʰ   December 21 at 2ʰ
January 6 at 1ʰ   January 21 at midnight
February 6 at 23ʰ   February 21 at 22ʰ
March 6 at 21ʰ   March 21 at 20ʰ

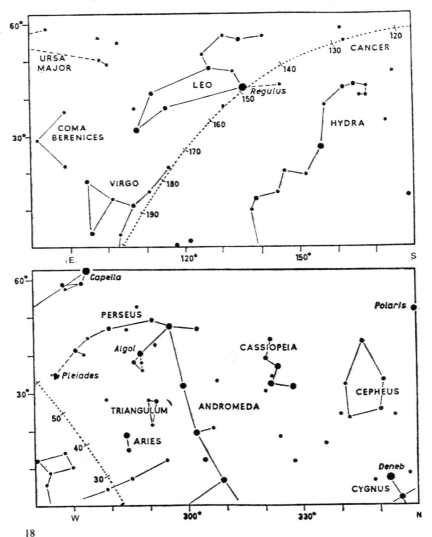

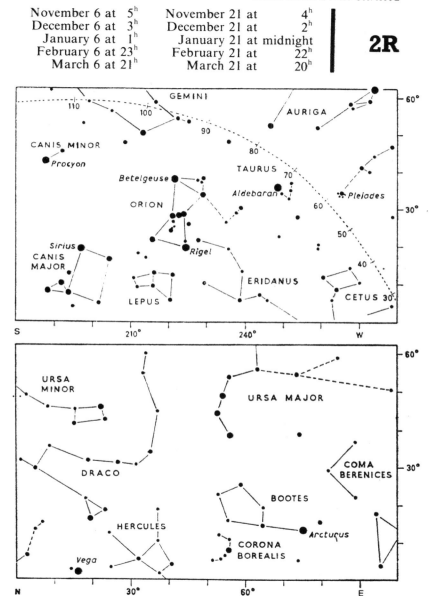

November 6 at 5^h November 21 at   4^h
December 6 at 3^h December 21 at   2^h
January 6 at 1^h January 21 at midnight
February 6 at 23^h February 21 at   22^h
March 6 at 21^h March 21 at   20^h

2R

GEMINI

110   100   90

AURIGA

80

CANIS MINOR

Procyon

TAURUS

70

Betelgeuse

Aldebaran

Pleiades

ORION

60

30°

50

Sirius

Rigel

CANIS
MAJOR

ERIDANUS

40

LEPUS

CETUS 30°

S   210°   240°   W

60°

URSA
MINOR

URSA MAJOR

COMA
BERENICES

30°

DRACO

BOOTES

HERCULES

Arcturus

CORONA
BOREALIS

Vega

N   30°   60°   E

19

3L

December 6 at 5^h December 21 at   4^h
January 6 at 3^h January 21 at   2^h
February 6 at 1^h February 21 at midnight
March 6 at 23^h March 21 at   22^h
April 6 at 21^h April 21 at   20^h

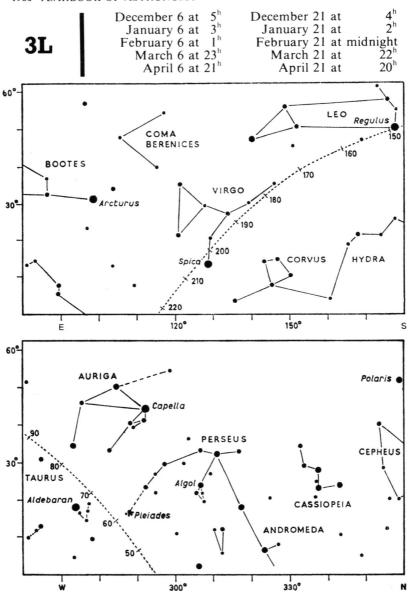

December 6 at 5^h	December 21 at 4^h
January 6 at 3^h	January 21 at 2^h
February 6 at 1^h	February 21 at midnight
March 6 at 23^h	March 21 at 22^h
April 6 at 21^h	April 21 at 20^h

3R

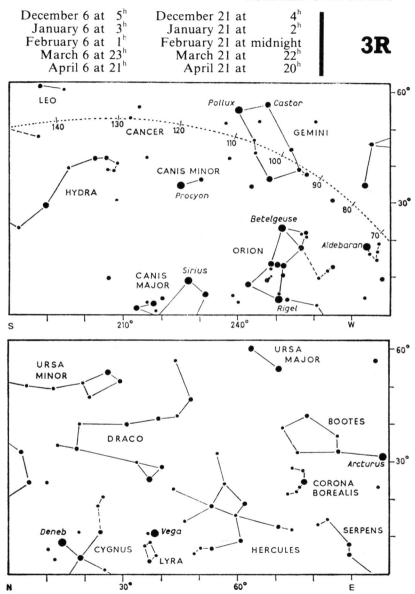

4L

January 6 at 5^h	January 21 at 4^h
February 6 at 3^h	February 21 at 2^h
March 6 at 1^h	March 21 at midnight
April 6 at 23^h	April 21 at 22^h
May 6 at 21^h	May 21 at 20^h

January 6 at 5ʰ January 21 at 4ʰ
February 6 at 3ʰ February 21 at 2ʰ
March 6 at 1ʰ March 21 at midnight
April 6 at 23ʰ April 21 at 22ʰ
May 6 at 21ʰ May 21 at 20ʰ

4R

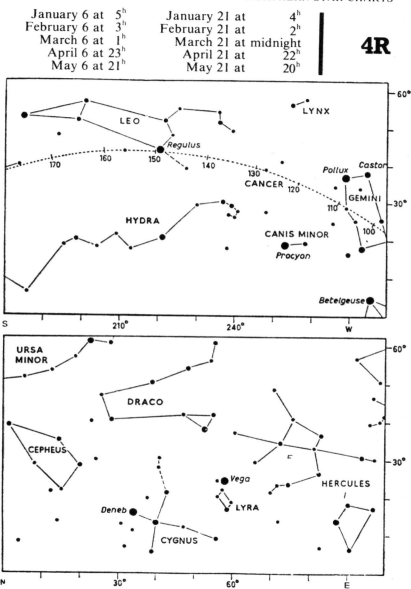

5L

January 6 at 7^h	January 21 at 6^h
February 6 at 5^h	February 21 at 4^h
March 6 at 3^h	March 21 at 2^h
April 6 at 1^h	April 21 at midnight
May 6 at 23^h	May 21 at 22^h

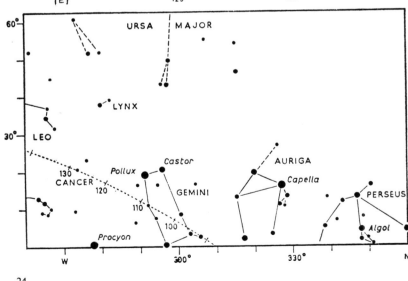

January 6 at 7ʰ   January 21 at   6ʰ
February 6 at 5ʰ   February 21 at   4ʰ
March 6 at 3ʰ   March 21 at   2ʰ
April 6 at 1ʰ   April 21 at midnight
May 6 at 23ʰ   May 21 at   22ʰ

5R

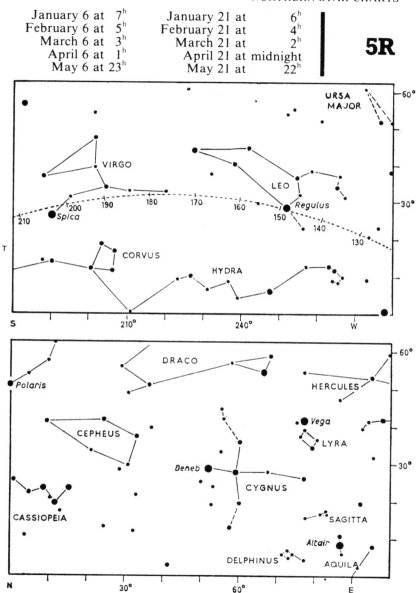

25

6L

March 6 at 5^h	March 21 at 4^h
April 6 at 3^h	April 21 at 2^h
May 6 at 1^h	May 21 at midnight
June 6 at 23^h	June 21 at 22^h
July 6 at 21^h	July 21 at 20^h

March 6 at 5ʰ   March 21 at 4ʰ
April 6 at 3ʰ   April 21 at 2ʰ
May 6 at 1ʰ   May 21 at midnight
June 6 at 23ʰ   June 21 at 22ʰ
July 6 at 21ʰ   July 21 at 20ʰ

6R

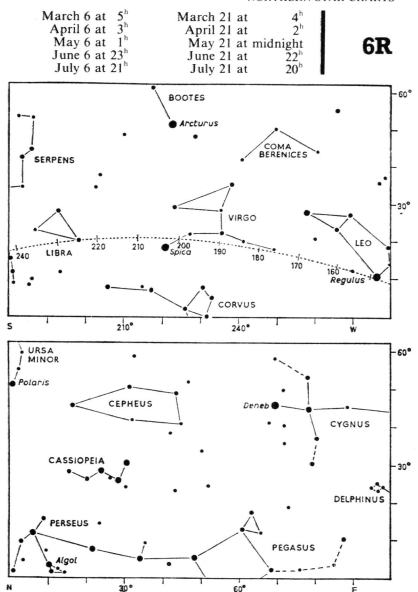

27

7L

May 6 at 3^h	May 21 at 2^h
June 6 at 1^h	June 21 at midnight
July 6 at 23^h	July 21 at 22^h
August 6 at 21^h	August 21 at 20^h
September 6 at 19^h	September 21 at 18^h

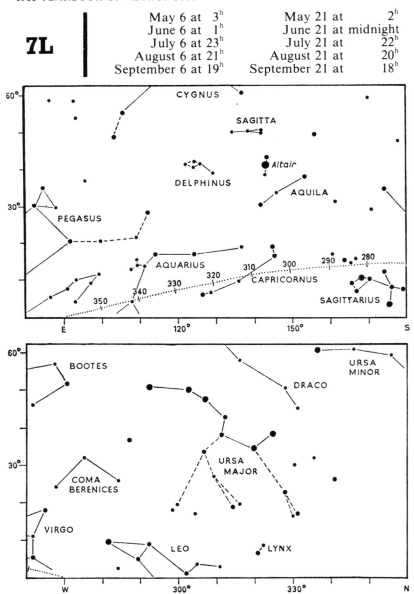

May 6 at 3ʰ | May 21 at 2ʰ
June 6 at 1ʰ | June 21 at midnight
July 6 at 23ʰ | July 21 at 22ʰ
August 6 at 21ʰ | August 21 at 20ʰ
September 6 at 19ʰ | September 21 at 18ʰ

7R

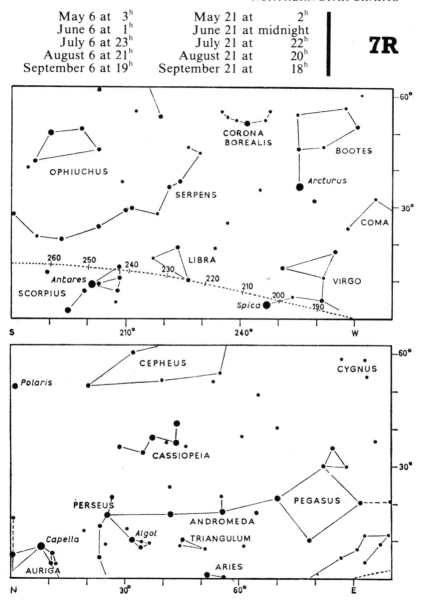

8L

July 6 at 1ʰ	July 21 at midnight
August 6 at 23ʰ	August 21 at 22ʰ
September 6 at 21ʰ	September 21 at 20ʰ
October 6 at 19ʰ	October 21 at 18ʰ
November 6 at 17ʰ	November 21 at 16ʰ

July 6 at 1ʰ   July 21 at midnight
August 6 at 23ʰ   August 21 at 22ʰ
September 6 at 21ʰ   September 21 at 20ʰ
October 6 at 19ʰ   October 21 at 18ʰ
November 6 at 17ʰ   November 21 at 16ʰ

8R

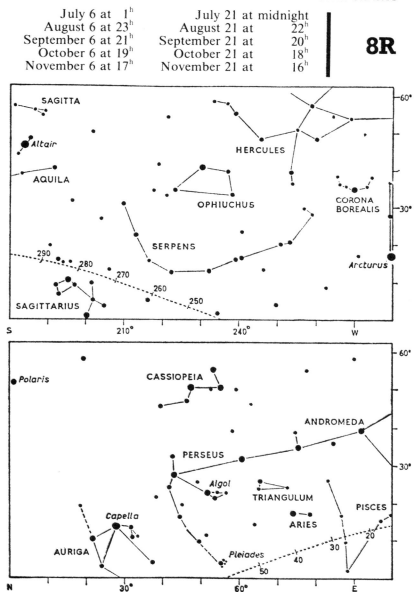

31

9L

August 6 at 1ʰ	August 21 at midnight
September 6 at 23ʰ	September 21 at 22ʰ
October 6 at 21ʰ	October 21 at 20ʰ
November 6 at 19ʰ	November 21 at 18ʰ
December 6 at 17ʰ	December 21 at 16ʰ

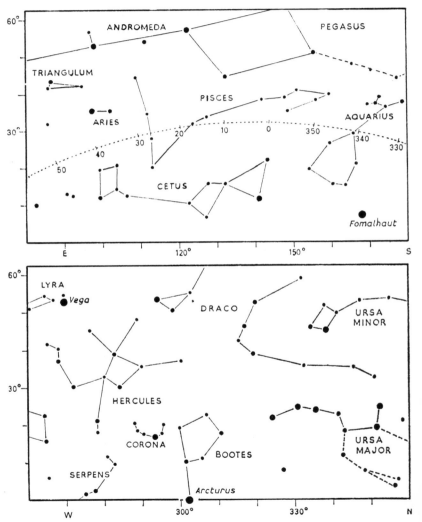

August 6 at 1ʰ   August 21 at midnight
September 6 at 23ʰ September 21 at   22ʰ
October 6 at 21ʰ   October 21 at   20ʰ
November 6 at 19ʰ   November 21 at   18ʰ
December 6 at 17ʰ   December 21 at   16ʰ

9R

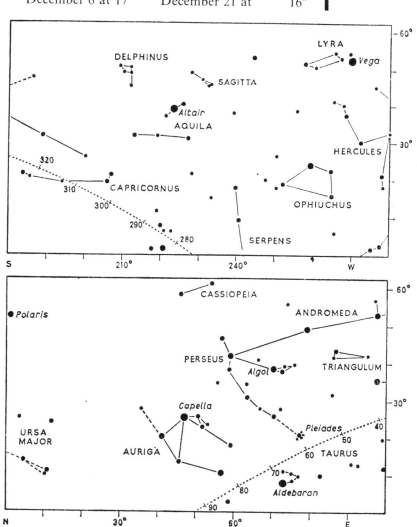

33

10L

August 6 at 3^h	August 21 at 2^h
September 6 at 1^h	September 21 at midnight
October 6 at 23^h	October 21 at 22^h
November 6 at 21^h	November 21 at 20^h
December 6 at 19^h	December 21 at 18^h

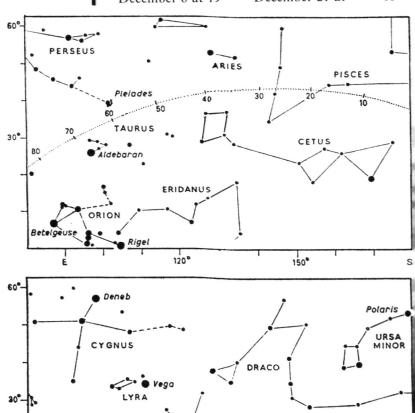

August 6 at 3^h	August 21 at 2^h
September 6 at 1^h	September 21 at midnight
October 6 at 23^h	October 21 at 22^h
November 6 at 21^h	November 21 at 20^h
December 6 at 19^h	December 21 at 18^h

10R

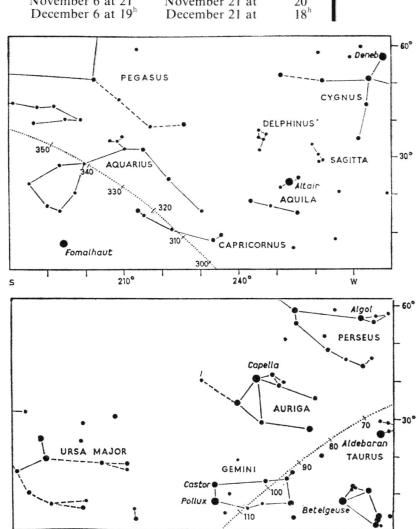

35

11L

September 6 at 3ʰ	September 21 at 2ʰ
October 6 at 1ʰ	October 21 at midnight
November 6 at 23ʰ	November 21 at 22ʰ
December 6 at 21ʰ	December 21 at 20ʰ
January 6 at 19ʰ	January 21 at 18ʰ

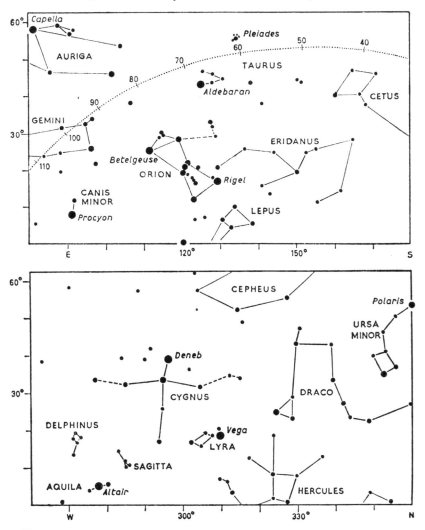

September 6 at 3^h   September 21 at   2^h
October 6 at 1^h   October 21 at midnight
November 6 at 23^h November 21 at   22^h
December 6 at 21^h December 21 at   20^h
January 6 at 19^h   January 21 at   18^h

11R

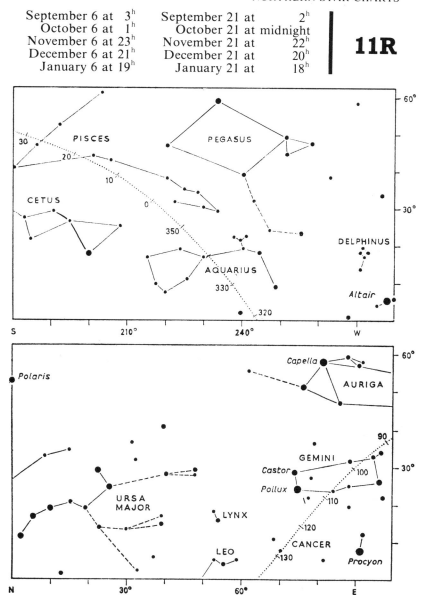

37

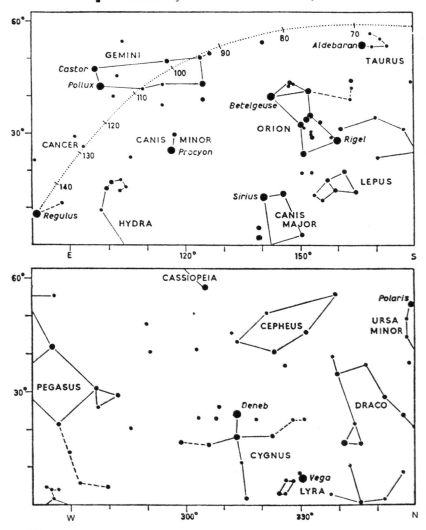

October 6 at 3^h	October 21 at 2^h
November 6 at 1^h	November 21 at midnight
December 6 at 23^h	December 21 at 22^h
January 6 at 21^h	January 21 at 20^h
February 6 at 19^h	February 21 at 18^h

12L

October 6 at   3ʰ   October 21 at   2ʰ
November 6 at   1ʰ   November 21 at midnight
December 6 at 23ʰ   December 21 at   22ʰ
January 6 at 21ʰ   January 21 at   20ʰ
February 6 at 19ʰ   February 21 at   18ʰ

12R

39

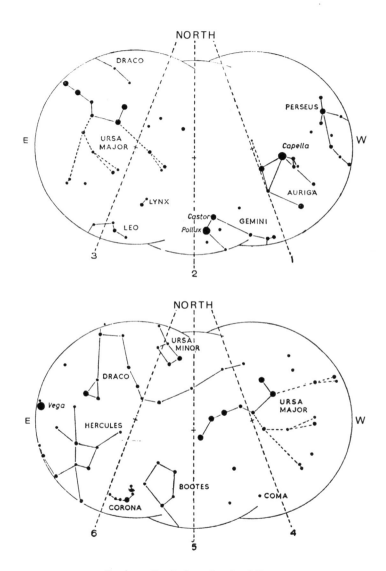

Northern Hemisphere Overhead Stars

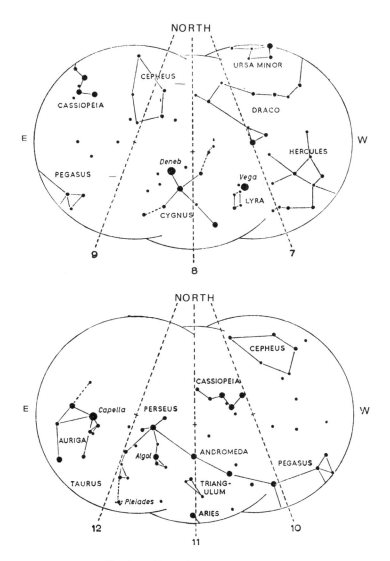

Northern Hemisphere Overhead Stars

1L

October 6 at 5^h	October 21 at 4^h
November 6 at 3^h	November 21 at 2^h
December 6 at 1^h	December 21 at midnight
January 6 at 23^h	January 21 at 22^h
February 6 at 21^h	February 21 at 20^h

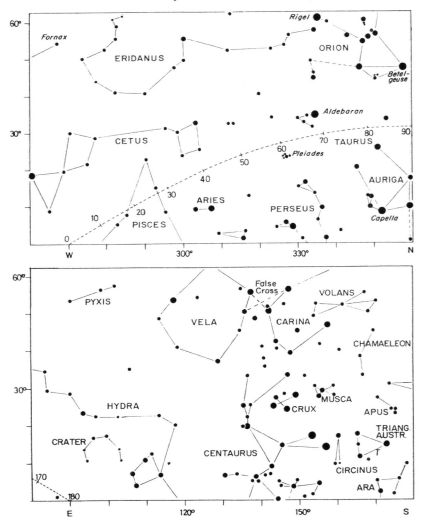

October 6 at   5^h October 21 at   4^h
November 6 at 3^h November 21 at   2^h
December 6 at 1^h December 21 at midnight
January 6 at 23^h January 21 at   22^h
February 6 at 21^h February 21 at   20^h

1R

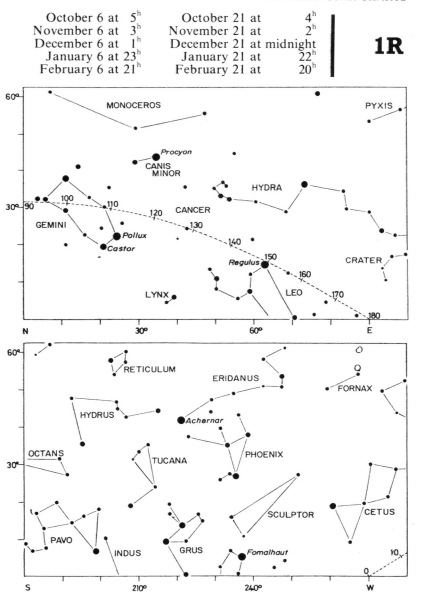

43

2L

November 6 at 5ʰ	November 21 at 4ʰ
December 6 at 3ʰ	December 21 at 2ʰ
January 6 at 1ʰ	January 21 at midnight
February 6 at 23ʰ	February 21 at 22ʰ
March 6 at 21ʰ	March 21 at 20ʰ

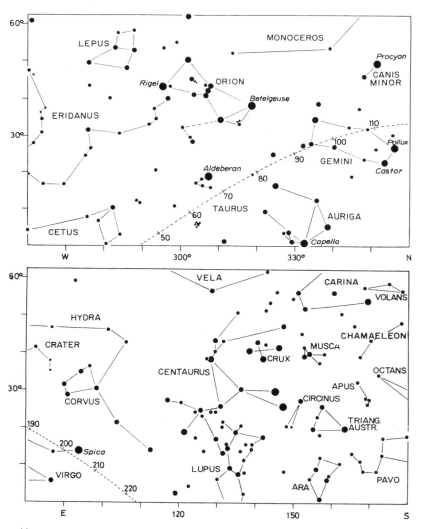

November 6 at 5^h	November 21 at 4^h	
December 6 at 3^h	December 21 at 2^h	**2R**
January 6 at 1^h	January 21 at midnight	
February 6 at 23^h	February 21 at 22^h	
March 6 at 21^h	March 21 at 20^h	

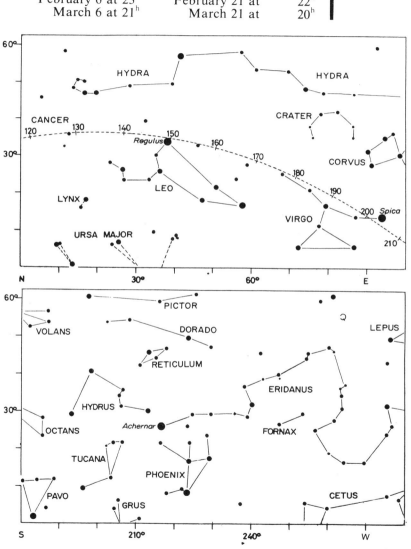

3L

January 6 at 3^h January 21 at 2^h
February 6 at 1^h February 21 at midnight
March 6 at 23^h March 21 at 22^h
April 6 at 21^h April 21 at 20^h
May 6 at 19^h May 21 at 18^h

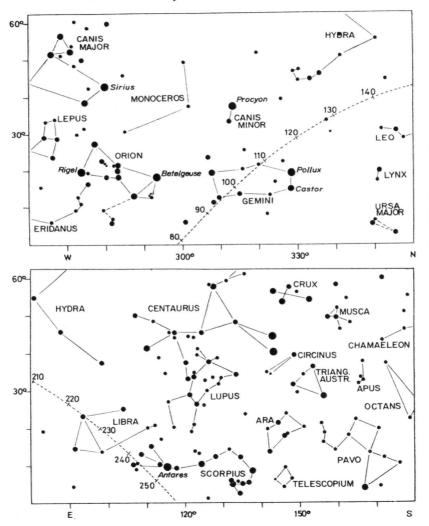

January 6 at 3ʰ	January 21 at 2ʰ
February 6 at 1ʰ	February 21 at midnight
March 6 at 23ʰ	March 21 at 22ʰ
April 6 at 21ʰ	April 21 at 20ʰ
May 6 at 19ʰ	May 21 at 18ʰ

3R

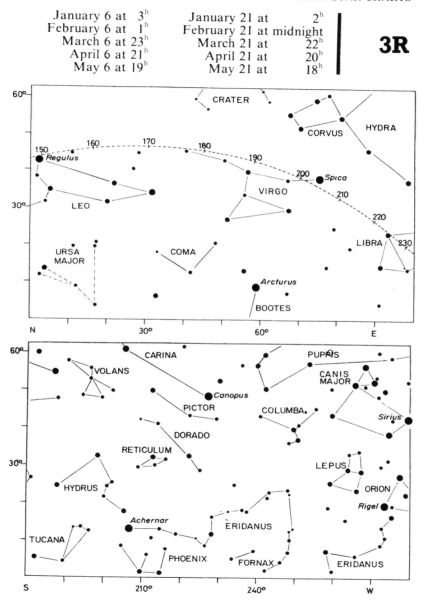

4L

February 6 at 3^h February 21 at 2^h
March 6 at 1^h March 21 at midnight
April 6 at 23^h April 21 at 22^h
May 6 at 21^h May 21 at 20^h
June 6 at 19^h June 21 at 18^h

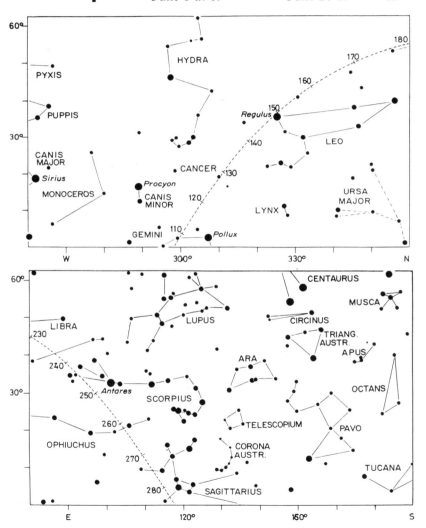

February 6 at 3ʰ   February 21 at 2ʰ
March 6 at 1ʰ   March 21 at midnight
April 6 at 23ʰ   April 21 at 22ʰ
May 6 at 21ʰ   May 21 at 20ʰ
June 6 at 19ʰ   June 21 at 18ʰ

4R

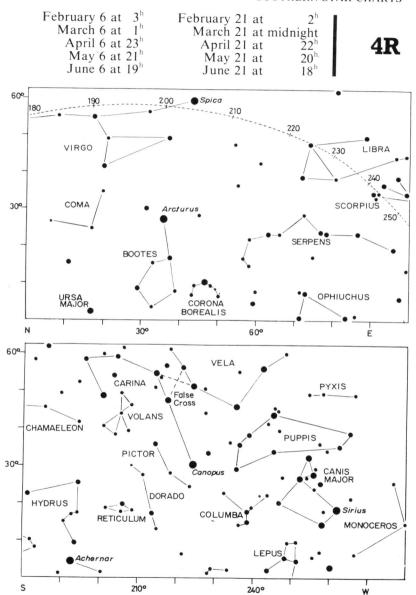

49

5L

March 6 at 3^h	March 21 at 2^h
April 6 at 1^h	April 21 at midnight
May 6 at 23^h	May 21 at 22^h
June 6 at 21^h	June 21 at 20^h
July 6 at 19^h	July 21 at 18^h

March 6 at 3^h March 21 at   2^h
April 6 at 1^h April 21 at midnight
May 6 at 23^h May 21 at   22^h
June 6 at 21^h June 21 at   20^h
July 6 at 19^h July 21 at   18^h

5R

6L

March 6 at 5ʰ	March 21 at 4ʰ
April 6 at 3ʰ	April 21 at 2ʰ
May 6 at 1ʰ	May 21 at midnight
June 6 at 23ʰ	June 21 at 22ʰ
July 6 at 21ʰ	July 21 at 20ʰ

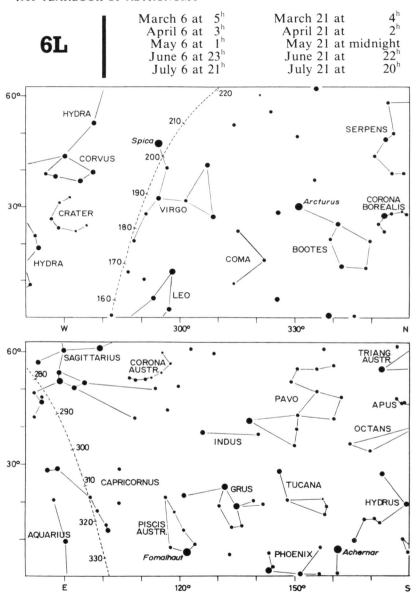

March 6 at 5^h March 21 at 4^h

April 6 at 3^h April 21 at 2^h

May 6 at 1^h May 21 at midnight

June 6 at 23^h June 21 at 22^h

July 6 at 21^h July 21 at 20^h

6R

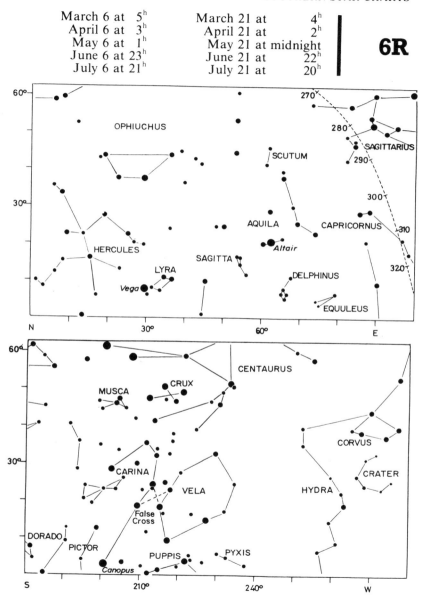

53

7L

April 6 at 5^h	April 21 at 4^h
May 6 at 3^h	May 21 at 2^h
June 6 at 1^h	June 21 at midnight
July 6 at 23^h	July 21 at 22^h
August 6 at 21^h	August 21 at 20^h

April 6 at 5^h   April 21 at 4^h
May 6 at 3^h   May 21 at 2^h
June 6 at 1^h   June 21 at midnight   **7R**
July 6 at 23^h   July 21 at 22^h
August 6 at 21^h   August 21 at 20^h

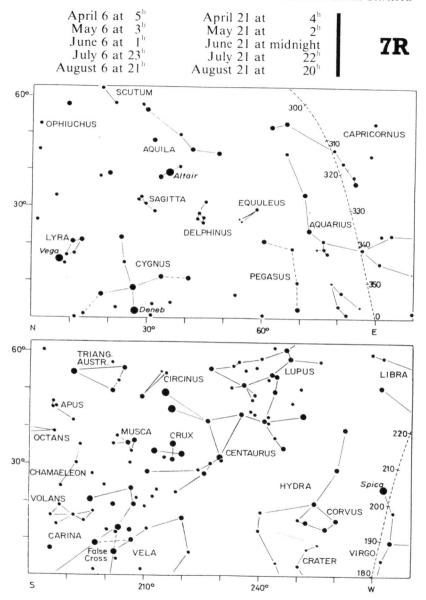

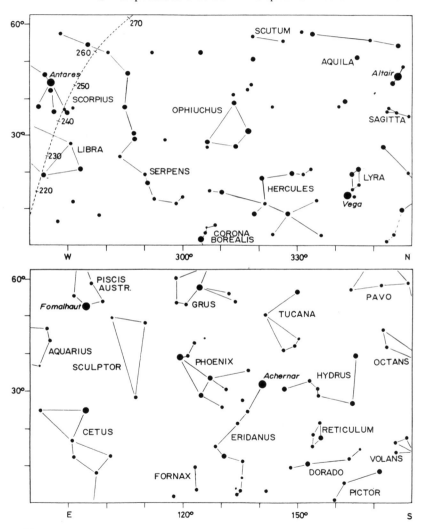

May 6 at 5ʰ	May 21 at 4ʰ
June 6 at 3ʰ	June 21 at 2ʰ
July 6 at 1ʰ	July 21 at midnight
August 6 at 23ʰ	August 21 at 22ʰ
September 6 at 21ʰ	September 21 at 20ʰ

8L

May 6 at 5ʰ   May 21 at   4ʰ
June 6 at 3ʰ   June 21 at   2ʰ
July 6 at 1ʰ   July 21 at midnight
August 6 at 23ʰ   August 21 at   22ʰ
September 6 at 21ʰ September 21 at 20ʰ

8R

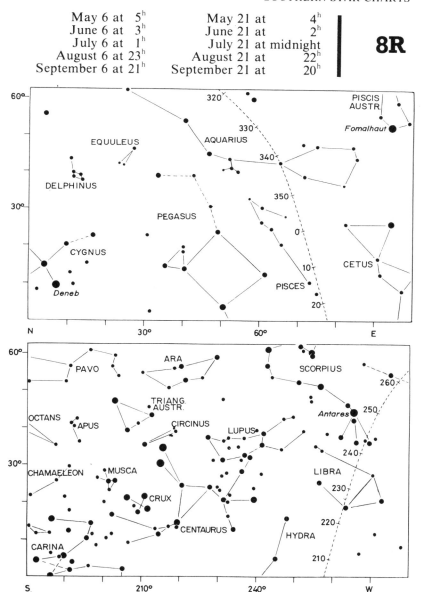

57

9L

June 6 at 5ʰ	June 21 at 4ʰ
July 6 at 3ʰ	July 21 at 2h
August 6 at 1ʰ	August 21 at midnight
September 6 at 23ʰ	September 21 at 22ʰ
October 6 at 21ʰ	October 21 at 20ʰ

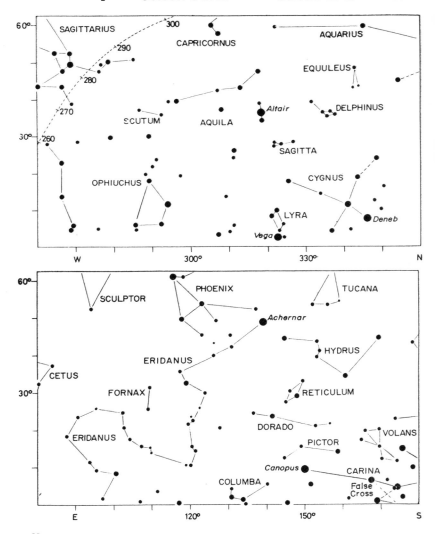

June 6 at 5^h	June 21 at 4^h	
July 6 at 3^h	July 21 at 2h	
August 6 at 1^h	August 21 at midnight	**9R**
September 6 at 23^h	September 21 at 22^h	
October 6 at 21^h	October 21 at 20^h	

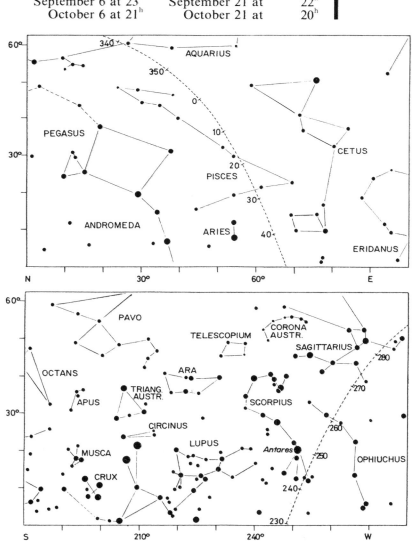

10L

July 6 at 5^h July 21 at 4^h
August 6 at 3^h August 21 at 2^h
September 6 at 1^h September 21 at midnight
October 6 at 23^h October 21 at 22^h
November 6 at 21^h November 21 at 20^h

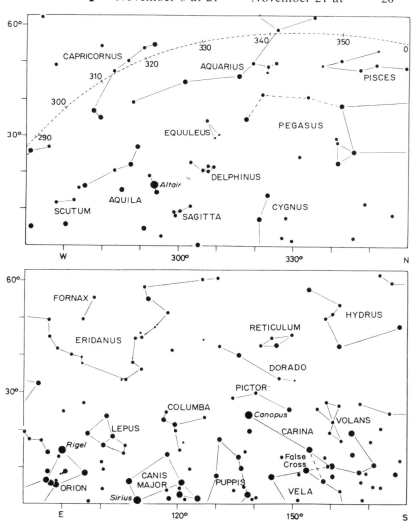

July 6 at 5[h]	July 21 at 4[h]
August 6 at 3[h]	August 21 at 2[h]
September 6 at 1[h]	September 21 at midnight
October 6 at 23[h]	October 21 at 22[h]
November 6 at 21[h]	November 21 at 20[h]

10R

11L

August 6 at 5^h August 21 at 4^h
September 6 at 3^h September 21 at 2^h
October 6 at 1^h October 21 at midnight
November 6 at 23^h November 21 at 22^h
December 6 at 21^h December 21 at 20^h

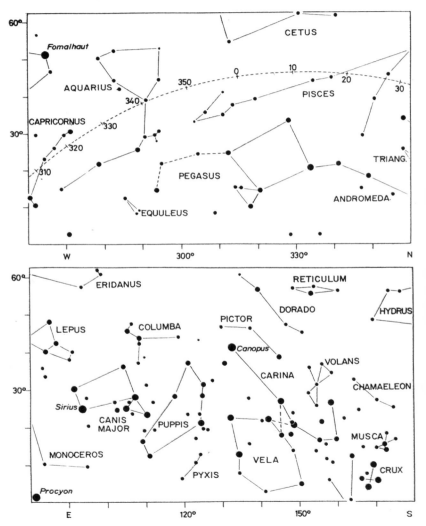

August 6 at 5ʰ
September 6 at 3ʰ
October 6 at 1ʰ
November 6 at 23ʰ
December 6 at 21ʰ

August 21 at 4ʰ
September 21 at 2ʰ
October 21 at midnight
November 21 at 22ʰ
December 21 at 20ʰ

11R

63

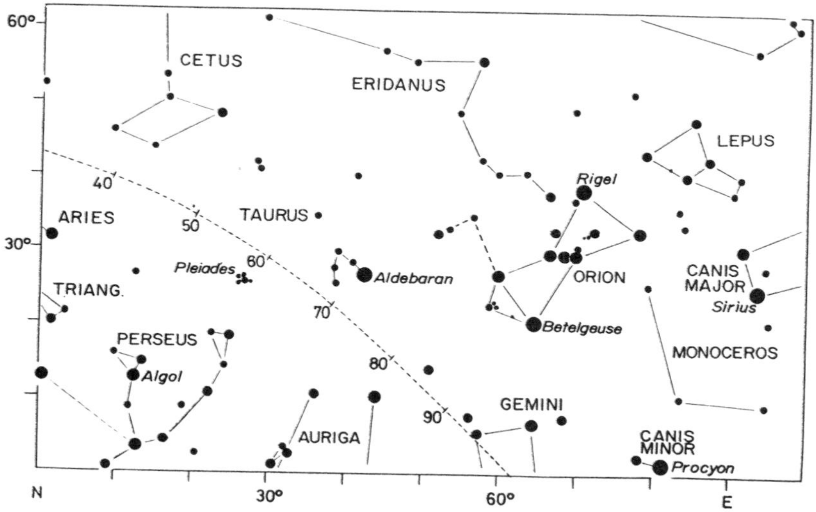

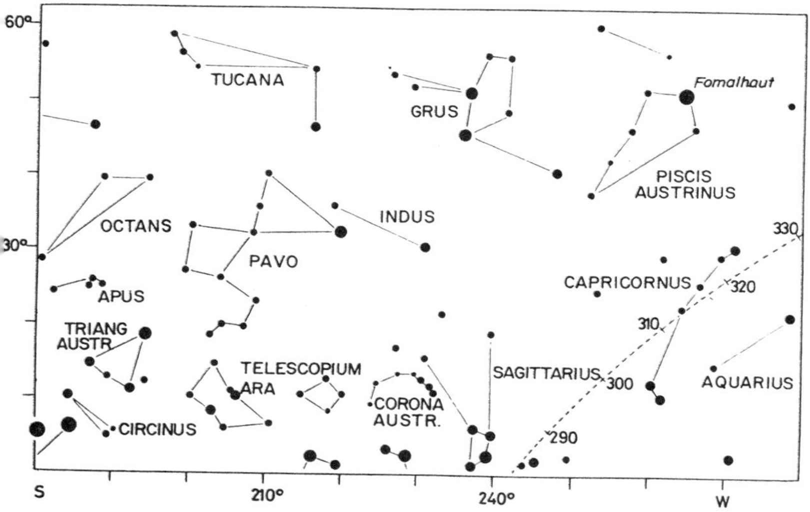

12L

September 6 at 5^h	September 21 at 4^h
October 6 at 3^h	October 21 at 2^h
November 6 at 1^h	November 21 at midnight
December 6 at 23^h	December 21 at 22^h
January 6 at 21^h	January 21 at 20^h

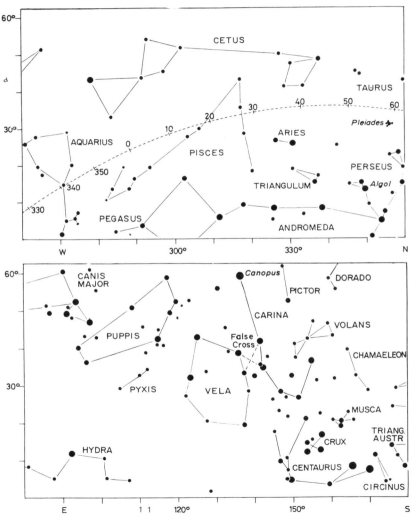

September 6 at 5^h September 21 at 4^h
October 6 at 3^h October 21 at 2^h
November 6 at 1^h November 21 at midnight
December 6 at 23^h December 21 at 22^h
January 6 at 21^h January 21 at 20^h

12R

65

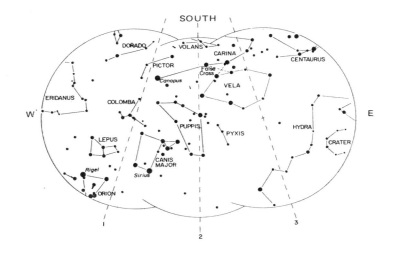

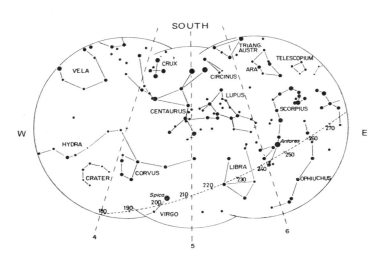

Southern Hemisphere Overhead Stars

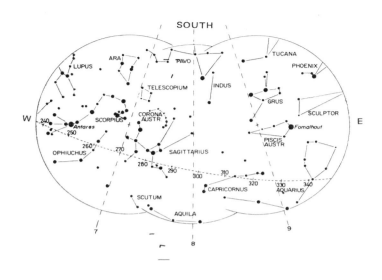

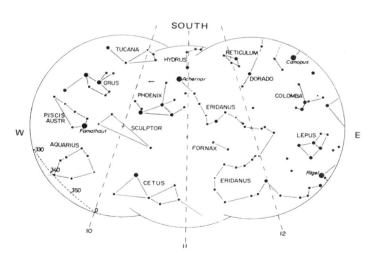

Southern Hemisphere Overhead Stars

The Planets and the Ecliptic

The paths of the planets about the Sun all lie close to the plane of the ecliptic, which is marked for us in the sky by the apparent path of the Sun among the stars, and is shown on the star charts by a broken line. The Moon and planets will always be found close to this line, never departing from it by more than about 7 degrees. Thus the planets are most favourably placed for observation when the ecliptic is well displayed, and this means that it should be as high in the sky as possible. This avoids the difficulty of finding a clear horizon, and also overcomes the problem of atmospheric absorption, which greatly reduces the light of the stars. Thus a star at an altitude of 10 degrees suffers a loss of 60 per cent of its light, which corresponds to a whole magnitude; at an altitude of only 4 degrees, the loss may amount to two magnitudes.

The position of the ecliptic in the sky is therefore of great importance, and since it is tilted at about 23½ degrees to the equator, it is only at certain times of the day or year that it is displayed to the best advantage. It will be realized that the Sun (and therefore the ecliptic) is at-its highest in the sky at noon in midsummer, and at its lowest at noon in midwinter. Allowing for the daily motion of the sky, these times lead to the fact that the ecliptic is highest at midnight in winter, at sunset in the spring, at noon in summer and at sunrise in the autumn. Hence these are the best times to see the planets. Thus, if Venus is an evening star, in the western sky after sunset, it will be seen to best advantage if this occurs in the spring, when the ecliptic is high in the sky and slopes down steeply to the north-west. This means that the planet is not only higher in the sky, but will remain for a much longer period above the horizon. For similar reasons, a morning star will be seen at its best on autumn mornings before sunrise, when the ecliptic is high in the east. The outer planets, which can come to opposition (i.e. opposite the Sun), are best seen when opposition occurs in the winter months, when the ecliptic is high in the sky at midnight.

The seasons are reversed in the Southern Hemisphere, spring beginning at the September Equinox, when the Sun crosses the Equator on its way south, summer begins at the December Solstice, when the Sun is highest in the southern sky, and so on. Thus, the times when the ecliptic is highest in the sky, and therefore best placed for observing the planets, may be summarised as follows:

	Midnight	Sunrise	Noon	Sunset
Northern lats.	December	September	June	March
Southern lats.	June	March	December	September

In addition to the daily rotation of the celestial sphere from east to west, the planets have a motion of their own among the stars. The apparent movement is generally *direct,* i.e. to the east, in the direction of increasing longitude, but for a certain period (which depends on the distance of the planet) this apparent motion is reversed. With the outer planets this *retrograde* motion occurs about the time of opposition. Owing to the different inclination of the orbits of these planets, the actual effect is to cause the apparent path to form a loop, or sometimes an S-shaped curve. The same effect is present in the motion of the inferior planets, Mercury and Venus, but it is not so obvious, since it always occurs at the time of inferior conjunction.

The inferior planets, Mercury and Venus, move in smaller orbits than that of the Earth, and so are always seen near the Sun. They are most obvious at the times of greatest angular distance from the Sun (greatest elongation), which may reach 28 degrees for Mercury, or 47 degrees for Venus. They are then seen as evening stars in the western sky after sunset (at eastern elongations) or as morning stars in the eastern sky before sunrise (at western elongations). The succession of phenomena, conjunctions and elongations, always follows the same order, but the intervals between them are not equal. Thus, if either planet is moving round the far side of its orbit its motion will be to the east, in the same direction in which the Sun appears to be moving. It therefore takes much longer for the planet to overtake the Sun—that is, to come to superior conjunction—than it does when moving round to inferior conjunction, between Sun and Earth. The intervals given in the following table are average values; they remain fairly constant in the case of Venus, which travels in an almost circular orbit. In the case of Mercury, however, conditions vary widely because of the great eccentricity and inclination of the planet's orbit.

		Mercury	Venus
Inferior conj.	to Elongation West	22 days	72 days
Elongation West	to Superior conj.	36 days	220 days
Superior conj.	to Elongation East	36 days	220 days
Elongation East	to Inferior conj.	22 days	72 days

The greatest brilliancy of Venus always occurs about 36 days before or after inferior conjunction. This will be about a month *after* greatest eastern elongation (as an evening star), or a month *before* greatest western elongation (as a morning star). No such rule can be given for Mercury, because its distance from the Earth and the Sun can vary over a wide range.

Mercury is not likely to be seen unless a clear horizon is available. It is seldom seen as much as 10 degrees above the horizon in the twilight sky in northern latitudes, but this figure is often exceeded in the Southern Hemisphere. This favourable condition arises because the maximum elongation of 28 degrees can occur only when the planet is at aphelion (farthest from the Sun), and this point lies well south of the Equator. Northern observers must be content with smaller elongations, which may be as little as 18 degrees at perihelion. In general, it may be said that the most favourable times for seeing Mercury as an evening star will be in spring, some days before greatest eastern elongation; in autumn, it may be seen as a morning star some days after greatest western elongation.

Venus is the brightest of the planets, and may be seen on occasions in broad daylight. Like Mercury, it is alternately a morning and an evening star, and will be highest in the sky when it is a morning star in autumn, or an evening star in spring. The phenomena of Venus given in the table above can occur only in the months of January, April, June, August and November, and it will be realized that they do not all lead to favourable apparitions of the planet. In fact, Venus is to be seen at its best as an evening star in northern latitudes when eastern elongation occurs in June. The planet is then well north of the Sun in the preceding spring months, and is a brilliant object in the evening sky over a long period. In the Southern Hemisphere a November elongation is best. For similar reasons, Venus gives a prolonged display as a morning star in the months following western elongation in November (in northern latitudes) or in June (in the Southern Hemisphere).

The superior planets, which travel in orbits larger than that of the Earth, differ from Mercury and Venus in that they can be seen opposite the Sun in the sky. The superior planets are morning stars after conjunction with the Sun, rising earlier each day until they come to opposition. They will then be nearest to the Earth (and therefore at their brightest), and will be on the meridian at midnight, due south in northern latitudes, but due north in the Southern Hemisphere. After opposition they are evening stars, setting earlier each evening until they set in the west with the Sun at the next conjunction. The change in brightness about the time of opposition is most noticeable in the case of Mars, whose distance from the Earth can vary considerably and rapidly. The other superior planets are at such great distances that there is very little change in brightness from one opposition to another. The effect of altitude is, however, of some importance, for at a December opposition in northern latitudes the planet will be among the stars of Taurus or Gemini, and can then be at an altitude of more than 60 degrees in southern England. At a summer opposition, when the planet is in Sagittarius, it may only rise to about 15 degrees above the southern horizon, and so makes a

71

less impressive appearance. In the Southern Hemisphere, the reverse conditions apply; a June opposition being the best, with the planet in Sagittarius at an altitude which can reach 78 degrees above the northern horizon.

Mars, whose orbit is appreciably eccentric, comes nearest to the Earth at an opposition at the end of August. It may then be brighter even than Jupiter, but rather low in the sky in Aquarius for northern observers, though very well placed for those in southern latitudes. These favourable oppositions occur every fifteen or seventeen years (1924, 1941, 1956, 1971) but in the Northern Hemisphere the planet is probably better seen at an opposition in the autumn or winter months, when it is higher in the sky. Oppositions of Mars occur at an average interval of 780 days, and during this time the planet makes a complete circuit of the sky.

Jupiter is always a bright planet, and comes to opposition a month later each year, having moved, roughly speaking, from one Zodiacal constellation to the next.

Saturn moves much more slowly than Jupiter, and may remain in the same constellation for several years. The brightness of Saturn depends on the aspect of its rings, as well as on the distance from Earth and Sun. The rings are now inclined towards the Earth and Sun at quite a small angle, and will be seen edge-on in 1980.

Uranus, Neptune, and *Pluto* are hardly likely to attract the attention of observers without adequate instruments, but some notes on their present positions in the sky will be found in the April, May and June Notes.

Phases of the Moon 1980

	New Moon				First Quarter				Full Moon				Last Quarter		
	d	h	m		d	h	m		d	h	m		d	h	m
								Jan.	2	09	02	Jan.	10	11	49
Jan.	17	21	19	Jan.	24	13	58	Feb.	1	02	21	Feb.	9	07	35
Feb.	16	08	51	Feb.	23	00	14	Mar.	1	21	00	Mar.	9	23	49
Mar.	16	18	56	Mar.	23	12	31	Mar.	31	15	14	Apr.	8	12	06
Apr.	15	03	46	Apr.	22	02	59	Apr.	30	07	35	May	7	20	51
May	14	12	00	May	21	19	16	May	29	21	28	June	6	02	53
June	12	20	38	June	20	12	32	June	28	09	02	July	5	07	27
July	12	06	46	July	20	05	51	July	27	18	54	Aug.	3	12	00
Aug.	10	19	09	Aug.	18	22	28	Aug.	26	03	42	Sept.	1	18	08
Sept.	9	10	00	Sept.	17	13	54	Sept.	24	12	08	Oct.	1	03	18
Oct.	9	02	50	Oct.	17	03	47	Oct.	23	20	52	Oct.	30	16	33
Nov.	7	20	43	Nov.	15	15	47	Nov.	22	06	39	Nov.	29	09	59
Dec.	7	14	35	Dec.	15	01	47	Dec.	21	18	08	Dec.	29	06	32

All times are G.M.T.

Reproduced, with permission, from data supplied by the Science Research Council.

The Planets in 1980

DATE		Venus	Mars	Jupiter	Saturn	Uranus	Neptune
January	6	318	165	160	177	234	261
	21	336	165	159	177	235	262
February	6	355	162	158	176	235	262
	21	13	157	156	175	236	262
March	6	29	152	154	174	236	263
	21	45	148	152	173	235	263
April	6	61	146	151	172	235	263
	21	75	147	150	171	235	263
May	6	86	151	150	170	234	262
	21	92	156	151	170	233	262
June	6	90	163	153	170	233	262
	21	81	170	155	171	232	261
July	6	77	178	157	172	232	261
	21	80	186	160	173	232	260
August	6	89	195	163	175	232	260
	21	102	205	166	176	232	260
September	6	118	215	169	178	232	260
	21	134	225	172	180	233	260
October	6	151	235	176	182	233	260
	21	169	246	179	184	234	261
November	6	188	258	182	185	235	261
	21	207	269	185	187	236	262
December	6	225	281	187	188	237	262
	21	244	292	189	189	238	263
Opposition:		——	Feb. 25	Feb. 24	Mar. 14	May 14	June 12
Conjunction:		June 15	——	Sep. 13	Sep. 23	Nov. 18	Dec. 14

Mercury moves so quickly among the stars that it is not possible to indicate its position on the star charts at a convenient interval. The monthly notes must be consulted for the best times at which the planet may be seen.

The positions of the other planets are given in the table on the previous page. This gives the apparent longitudes on dates which correspond to those of the star charts, and the position of the planet may at once be found near the ecliptic at the given longitude.

Examples:

(1)   *Where may the planet Saturn be found at the end of January?*

The table opposite gives the longitude of Saturn at this time as 177 degrees. For northern observers the upper diagram of Northern Chart 2L shows that Saturn may be seen to the south of east at midnight, below the figure of Leo. The planet will be due south (Chart 4R) at 4^h.

In the southern hemisphere Chart 2R (Southern) is appropriate, and the planet may be seen to the north of east at midnight, to the right of the figure of Leo. It will be seen that Saturn would be due north on Chart 4L, but this would be at 4^h (see table on page 14) and dawn will then have broken in latitude 35 degrees south.

(2)   *Identify the two bright planets seen near Regulus in the late evening in the first week of April.*

Northern Star Charts 2L or 3L show the position of the ecliptic at this time, and the longitude of Regulus is seen to be about 150 degrees. The table opposite identifies the planets as Mars (146 degrees) and Jupiter (151 degrees). Since longitudes increase from west to east, Mars must be the planet to the west (right) of Regulus, while Jupiter is to the east (left) of the star. In the southern hemisphere Star Charts 3R and 4L (southern) will apply, and lead to the same conclusions. The difference here is that the figure of Leo is inverted and will be seen in the north, with Mars to the left of Regulus, and Jupiter to the right.

Some Events in 1980

ECLIPSES
In 1980 there will be two eclipses, both of the Sun.
>16 February—a total eclipse of the Sun, visible in Africa and southern Asia.
>10 August—an annular eclipse of the Sun, visible in Hawaii, the West Indies and South America.

THE PLANETS
>*Mercury* may best be seen in northern latitudes at eastern elongation (evening star) on 19 February, and at western elongation (morning star) on 19 November. In the Southern Hemisphere the best dates are 2 April (morning star) and 11 October (evening star).
>
>*Venus* will be an evening star until inferior conjunction on 15 June, and a morning star for the rest of the year. Greatest eastern elongation on 5 April, greatest western elongation on 24 August.
>
>*Mars* is at opposition on 25 February in Leo, but will then be at aphelion.
>
>*Jupiter* is at opposition on 24 February, also in Leo and close to Mars. Conjunction occurs on 13 September.
>
>*Saturn* is at opposition on 14 March on the borders of Leo and Virgo, and will be in conjunction on 23 September.
>
>*Uranus* is at opposition on 14 May in Libra.
>
>*Neptune* is at opposition on 12 June and is still in Ophiuchus.
>
>*Pluto* is at opposition on 10 April on the borders of Virgo and Boötes.

January

Full Moon: 2 January   *New Moon:* 17 January

Earth is at perihelion (nearest to the Sun) on 3 January, when its distance will be 91.4 million miles (147.1 million km).

Mercury is in superior conjunction on 21 January, and will not be visible during the month.

Venus is an evening star, moving out to greatest elongation in April. In the northern hemisphere the planet sets in the south-west two to three hours after the Sun, and will be a brilliant object in the evening sky throughout the spring months. Venus is not so well placed for observers in southern latitudes, and will set about two hours after sunset throughout the early months of the year. Magnitude –3.4 to –3.5

Mars rises in the east in the late evening, and will be found south of the figure of Leo. The planet reaches a stationary point on 17 January, and then begins its retrograde movement, growing brighter as it approaches opposition (magnitude +0.4 to –0.6). Mars was in conjunction with Jupiter in December, and there will be two more conjunctions with that planet in March and May. The diagram below shows how the loop in the apparent path of Mars overlaps that of Jupiter; the March and May conjunctions are marked.

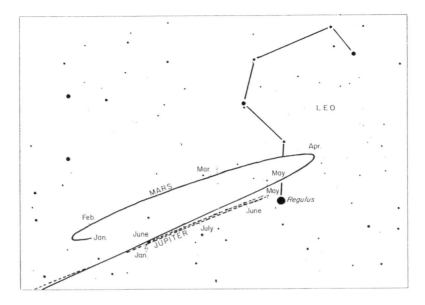

Mars and Jupiter

Jupiter will be seen a few degrees to the west of Mars, but is much brighter (magnitude –1.8 to –2.0), moving retrograde as it approaches opposition in February. A feature of the early months of 1980 is the proximity of the three bright planets, Mars, Jupiter and Saturn. In the southern hemisphere all three planets will be seen in the north at dawn, but in northern latitudes, with their longer nights, the planets will be well to the west of the south point by daybreak. The behaviour of the four great satellites of Jupiter is always of interest, even with the simplest of optical aids, and during the year all four undergo eclipses and occultations.

Saturn reaches a stationary point on 8 January and will be found to the east of Mars and Jupiter, and south of Denebola (the tail of Leo). See diagram in the April notes. Saturn rises just before midnight for observers in both hemispheres, but the planet is not a very brilliant object at this opposition. The rings are presented almost edge-on to the Earth and Sun, with the Sun on the south side of the rings, and the Earth on the north. As a result the rings are invisible, at least in small telescopes, and they provide no light whatever, so that the magnitude of the planet is only $+ 1.2$ to $+ 0.9$.

THE QUADRANTIDS

The Quadrantid meteor shower, with its maximum during the night of January 3-4, is one of the most interesting of the annual displays. It is named after the rejected constellation Quadrans Muralis (the Mural Quadrant), not far from Ursa Major. In fact, this is the only reason why Quadrans is remembered today.

Most meteor showers have a 'build-up' period before maximum, and a period of decay afterwards, but this is only roughly true for the Quadrantids, where the maximum is remarkably short and sharp. The Zenithal Hourly Rate, or Z.H.R., may be as high as 250. The Z.H.R. is the probable hourly rate of meteors which would be seen by an observer working under ideal conditions, with the radiant at the zenith or overhead point of the sky. Obviously, these conditions are virtually never achieved, but the Z.H.R. is a good guide to the richness of the shower. It is seldom as much as 60 for the Perseids, but of course the Perseid shower extends over more than a fortnight from late July to mid-August. The average Z.H.R. of the Quadrantids is given as 110, and anything in excess of 200 is exceptional, but it does happen sometimes. The Quadrantid meteors are in general white or bluish-white, and leave fine trails, so that they are well worth watching. Because the radiant is in the far north of the sky, southern-hemisphere observers are at a marked disadvantage, and it is indeed true to say that most of the major meteor showers seen each year have their radiants well north of the celestial equator.

It must however be added that not all meteors are members of showers. Sporadic meteors may appear from any direction at any moment.

JUNO AND THE LARGEST ASTEROIDS

On 13 January Juno, No. 3 in the list of minor planets or asteroids, comes to opposition. It lies in Canis Minor, not far from Procyon, and since its magnitude is only 7.7 it looks exactly like a faint star. The only way to identify it is by checking its motion from night to night. The movement is easily noted from one night to the next.

Juno was discovered on 1 September 1804 by the German astronomer Harding. Previously only Ceres and Pallas had been known. Vesta was added to the list in 1807, but the fifth member of the swarm, Astræa, was not discovered until 1845, so that for several decades only the four 'originals' were known. Ceres, Pallas, and Vesta are the largest members of the entire group, but Juno is smaller, and comes only equal eleventh in order of size. The list of the largest asteroids is as follows:

Asteroid	Diameter	
	miles	km
Ceres (1)	623	1003
Pallas (2)	378	608
Vesta (4)	334	538
Hygeia (10)	280	450
Euphrosyne (31)	230	370
Interamnia (704)	218	350
Davida (511)	201	323
Cybele (65)	192	309
Europa (52)	180	289
Patientia (451)	172	276
Psyche (16)	155	250
Doris (48)	155	250
Undina (92)	155	250
Juno (3)	155	250

However, Juno comes fifth in order of mean opposition apparent magnitude.
The list here is:

Asteroid	Mean opposition magnitude
Vesta (4)	6.4
Ceres (1)	7.3
Pallas (2)	7.5
Iris (7)	7.8
Juno (3)	8.1
Hebe (6)	8.3

Juno has a rotation period of 7.22 hours. The escape velocity is, of course, so low that no trace of atmosphere can exist. The albedo or reflecting power is only 15%, about the same as that of Mars.

Asteroid diameters have never been easy to measure, and the values given here are only approximate, but they are certainly of the right order. The origin of the asteroids is still a matter for debate. It has been suggested that they represent the débris of a former planet (or planets) which broke up, but most authorities now prefer the theory that the asteroidal material was never part of a larger body. It is also significant that all the asteroids combined would not add up to a single satellite the mass of our Moon.

GEORGES VAN BIESBROECK

21 January is the centenary of the birth of the famous Belgian astronomer Georges Achille van Biesbroeck. He was assistant successively at Heidelberg, Potsdam, and Uccle Observatories before emigrating to America in 1915, where he became assistant at the Yerkes Observatory. He discovered several comets, and carried out valuable studies of double stars.

February

Full Moon: 1 February   *New Moon:* 16 February

Mercury is at greatest eastern elongation on 19 February (18 degrees) and is at perihelion on the same day. This is a favourable opportunity to see Mercury as an evening star, low in the south-west at sunset. The diagram shows the changes in

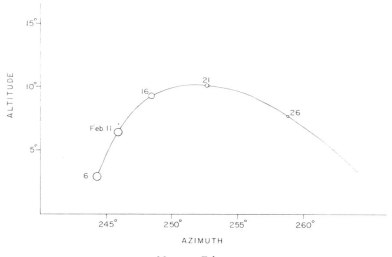

Mercury, February

altitude and azimuth (true bearing from the north through east, south and west) of Mercury on successive evenings when the Sun is six degrees below the horizon in latitude 52 degrees north (this is about 35 minutes after sunset). The changes in brightness are roughly indicated by the size of the circles, and it will be seen that Mercury is brightest *before* the date of eastern elongation. Conditions are less favourable in the southern hemisphere, but the planet should be bright enough to be glimpsed in the west in the middle of February.

Venus is a brilliant evening star and will be seen in the western sky for three hours or more after sunset in the northern hemisphere. Magnitude –3.5 to –3.7. In southern latitudes the planet sets less than two hours after the Sun.

Mars is at opposition on 25 February and is at aphelion on the same day at a distance from the Earth of 63 million miles (101 million km), and 155 million miles from the Sun. This is therefore not a very favourable opposition, the planet being less bright than it can be at an August opposition (at perihelion), when the distance from the Earth may be only 35 million miles, and the magnitude reaches –2.5. In 1980 the magnitude of Mars is only –1.0 at opposition.

Jupiter is at opposition on 24 February at 18^h at a distance of 409 million miles (659 million km). The oppositions of Mars and Jupiter are only 12 hours apart, and it is this fact that accounts for the triple conjunction. The brightness of Jupiter at opposition does not greatly vary. It may reach a magnitude of –2.5 when opposition occurs at perihelion (i.e., opposition in October), but may be half a magnitude less than this at aphelion. The magnitude in 1980 is –2.1 at opposition.

Saturn is moving retrograde in Virgo, and in mid-February will be seen about 10 degrees south of the second magnitude star Denebola at the tail of the Lion. Saturn rises in the late evening at the beginning of the month, and grows brighter as it approaches opposition. Magnitude +1.0 to +0.9.

A total eclipse of the Sun on 16 February will be visible in the Pacific Ocean and South America. Visible as a partial eclipse in

Hawaii, the south-western parts of the U.S.A., the West Indies and South America. For details see page 127.

MARS AT APHELION

Since Mars comes to opposition and aphelion on the same day this month, the opposition distance is almost as great as it can ever be. Since the planet lies in Leo, there is some compensation for northern-hemisphere observers, but for those astronomers who live in the southern hemisphere conditions are particularly unfavourable – small apparent diameter (only 13″.8) and relatively low altitude. Over the next decade, however, the opposition magnitude will brighten as the opposition distance becomes smaller. Oppositions between now and 1990 are:

Opposition Date	Apparent diameter seconds of arc	Magnitude	Constellation
1982 March 31	14.7	−1.2	Virgo
1984 May 11	17.5	−1.8	Libra
1986 July 10	23.1	−2.4	Sagittarius
1988 September 28	23.7	−2.6	Pisces
1990 November 27	17.9	−1.7	Taurus

At the most favourable oppositions, the apparent diameter may reach 25″.7, so that 1988 will be a good time for northern hemisphere observers.

PLANETARY CONJUNCTIONS

Though Mars and Jupiter are comparatively close together during the early part of 1980, there are no really close conjunctions and certainly no occultation of Jupiter by Mars. Mutual planetary occultations are, in fact, extremely rare. Since 1800 there have been only two which have been easily observable: on 9 December 1808 (Saturn occulted by Mercury) and 3 January 1818 (Jupiter occulted by Venus). None occur during the twentieth century, and the next will be on 22 November 2065 (Jupiter occulted by Venus), 11 August 2079 (Mars occulted by Mercury), and 27 October 2088 and 7 April 2094 (Jupiter occulted by Mercury). The last occasion when a first-magnitude star was occulted by a planet was on 7 July 1959, when Venus passed in front of Regulus. No more first-magnitude stars will be occulted by planets before the end of the century, but on 17 November 1981 Venus will occult the second-magnitude star Sigma Sagittarii.

THIS MONTH'S TOTAL ECLIPSE

At the time of this month's eclipse the Sun should be almost at the peak of its cycle of activity, and this will affect the shape of the corona. Generally, the corona is more symmetrical at spot-maximum than at spot-minimum, when there are polar 'streamers' – as was first recognized more than a hundred years ago.

Total solar eclipses have always been regarded as of extreme importance, because they have provided the only chances of studying the corona in detail. This is still true in the main, though observations from manned space-stations (such as Skylab) have altered the situation considerably. After the present eclipse, we must wait for 31 July 1981 (Russia, North Pacific area), 11 June 1983 (Indian Ocean, Pacific area) and 22/3 November 1984 (East Indies, South Pacific). No total eclipse can last for as long as 8 minutes. There will be an interesting eclipse on 3 October 1986; it will be annular along most of the track, but total for about one-tenth of a second over a small area in the North Atlantic. One may expect that many ships will be crowded into that particular region!

REGULUS

Regulus, or Alpha Leonis, is the faintest star usually reckoned as being of the first magnitude. The actual magnitude is 1.36, slightly fainter than Deneb in Cygnus; it comes 21st in order of brilliancy. The absolute magnitude (that is to say, the apparent magnitude that it would have if seen from the standard distance of 10 parsecs, or 32.6 light-years) is –0.7. Regulus is a hot star; the spectral type is B7, so that the surface temperature is over 12,000 degrees Centigrade. The distance has been given as 84 light-years, and Regulus is approximately 170 times as luminous as the Sun. It lies in the curved line of stars which has been nicknamed the Sickle. Regulus has a northerly declination of just over 12 degrees, so that it is visible from every inhabited continent. Only from parts of Antarctica does it remain permanently below the horizon.

March

Full Moon: 1 and 31 March   *New Moon:* 16 March

Summer Time in Great Britain and Northern Ireland commences 16 March.

Equinox: 20 March

Mercury is in inferior conjunction on 6 March, and after this it becomes a morning star. The planet is south of the equator and is not likely to be seen in northern latitudes. In the southern hemisphere Mercury will be well above the horizon in the east from mid-March onwards, but the planet is approaching aphelion and is not very bright.

Venus continues to make a splendid appearance as an evening star, and is growing brighter (magnitude –3.7 to –4.0). At northern stations the planet sets more than four hours after the Sun, but at 35 degrees south it will be visible for less than two hours after sunset.

Mars is an evening star in Leo, moving retrograde but fading rapidly as its distance from the Earth increases (magnitude –1.0 to –0.3). This rapid change of brightness at opposition is characteristic of Mars. On 2 March at 19^h, Mars passes about 3 degrees north of Jupiter, the second of the three conjunctions of these two planets. (See diagram in the January notes). Mars is also in conjunction with Regulus – also for the second time – on 17 March. The planet is visible for most of the night, setting about an hour before the Sun in northern latitudes, but in the southern hemisphere it is less favourably placed, being north of the equator, and it sets well before dawn.

Jupiter is to be seen in Leo to the east of Regulus and not far from Mars. The conjunction of the two planets on March 2 is mentioned above, and by the end of the month Mars will be about five degrees to the west and north of Jupiter. Jupiter is much brighter than Mars (magnitude −2.1 to −2.0), and is always easily recognized. The rapidly changing configuration of these planets at the foot of the Sickle of Leo will continue to be of interest for some weeks.

Saturn is at opposition on 14 March at a distance of 785 million miles (1264 million km), and is then of magnitude +0.9. The planet can be 1½ magnitudes brighter than this when the rings are wide open, as in 1973. On 3 March the plane of the rings passes through the Sun, and from this date the Sun will be on the north side of the rings. For a few days Sun and Earth will be on the same side of the rings, but on 12 March the ring-plane passes through the Earth for the second time This state of affairs will continue until the end of July, when the ring-plane will again pass through the Earth, and the rings will gradually come into view once more. On these occasions of ring-passage through Earth and Sun, the satellites of Saturn undergo eclipses and occultations similar to the phenomena of Jupiter's satellites. Although the largest satellite, Titan can be seen with quite a small instrument, the others are much more difficult to observe.

A penumbral eclipse of the Moon on 1 March is the first of three during the year. These penumbral eclipses attract little attention since the change of light is very small. See notes on page 128.

THE SATELLITES OF SATURN

This year, when the rings of Saturn are virtually edge on, conditions are exceptionally favourable for observing the satellites of the planet. Of these, Titan is a planet-sized world, larger than Mercury and not a great deal smaller than Mars. It has a relatively dense atmosphere (the surface pressure is at least ten times as great as that of the atmosphere of Mars), and it is a prime target for the first Voyager probe, which will by-pass Saturn next November. The escape velocity of Titan is about 4 miles per second, and there seems no doubt that the atmosphere contains clouds.

If Voyager 1 is successful in surveying Titan as well as Saturn, then Voyager 2, due to make its pass during 1981, will be able to move outward toward encounters with Uranus (1986) and Neptune (1989 or 1990). If, however, Titan is not surveyed by Voyager 1, then Voyager 2 will have to make the attempt, and 'miss out' on Uranus and Neptune. Understandably, astronomers are therefore particularly anxious that Voyager 1 should succeed.

Only two other satellites of Saturn are thought to have escape velocities of over one mile per second: Dione (1.4) and Iapetus (1.1) Dione is less than a thousand miles in diameter, and is therefore considerably smaller than our Moon, but it seems to be comparatively dense. Iapetus has always been very much of a puzzle. It is variable over a wide range, and is always brighter when west of Saturn than when to the east of the planet – as has been known ever since 1672, following a series of observations by G.D. Cassini, who had discovered Iapetus in the previous year. Either Iapetus must be irregular in shape, or else the two hemispheres must be very unequal in albedo.

Of the remaining satellites, Rhea is a fairly easy object with a small telescope, but the rest are more elusive. Considerable uncertainty attaches to Janus, which is the closest to Saturn of the entire retinue, and was discovered in 1966 by Audouin Dollfus (the last occasion on which the rings were edgewise-on). Janus is virtually unobservable except when the rings are out of view, and some astronomers even doubt its existence. Alternatively, it has been suggested that there may be several small inner satellites moving closer to Saturn than Mimas, which was discovered by William Herschel as long ago as 1789. Observations carried out this year (and by the various probes) may clear the matter up.

Conditions are also good for observing features on the disk of Saturn itself. The surface is less active than that of Jupiter, but there are occasional outbreaks – as in August 1933, when a prominent white spot was discovered by W. T. Hay and remained identifiable for over a month.

CANOPUS

European observers never fail to bemoan the fact that some of the most glorious of the southern stars remain permanently out of view. One of these is Canopus, leader of Carina (the Keel of the now-dismembered Ship Argo). Canopus has a southerly declination of 52½ degrees, so that to be visible the observer must go south of latitude 38 degrees North. One of the first practical proofs of the spherical shape of the Earth, as given by Aristotle, was that Canopus can be seen from Alexandria, but never from Athens!

Canopus is of spectral type FO. F-type stars are usually said to be yellowish, but to the observer with average eyesight Canopus will be described as colourless.

Apart from Sirius, it is the most brilliant star in the sky; the apparent magnitude is –0.72. Unlike Sirius, it is a true celestial searchlight. Estimates of its distance and luminosity vary widely, but at least Canopus must be thousands of times more powerful than the Sun.

To observers in Australia and South Africa, Canopus is not far from the zenith during February evenings, and it cannot be mistaken. Yet during the 1830s and 1840s it was rivalled, and sometimes surpassed, by the extraordinary variable Eta Carinæ, which lies not far from it. Today, however, Eta Carinæ is below naked-eye visibility, and has been so for a century or so. It was described in detail by David Allen in the 1979 issue of the *Yearbook*.

April

New Moon: 15 April   *Full Moon:* 30 April

Mercury is at greatest western elongation (28 degrees) on 2 April and is a morning star well placed for observers in southern latitudes. It will be found in the eastern sky at dawn and grows brighter during the month. By the end of April it will still be well above the horizon to the north of west in the dawn sky.

Venus is at greatest eastern elongation (46 degrees) on 5 April and is a brilliant object in the western sky. Venus is still setting later each evening and by the end of April it will be nearly midnight before it sets to the north of west (northern hemisphere). In southern latitudes the planet continues to set about two hours after the Sun. On 15 April Venus reaches its greatest northern latitude, and on the same day the planet passes 9 degrees north of Aldebaran. Since Aldebaran is 5½ degrees south of the ecliptic, Venus must be 3½ degrees north of it. This figure may be greatly exceeded when Venus is nearest the Earth at inferior conjunction, and on some of these occasions Venus may be more than 8 degrees north or south of the Sun.

Mars is still visible for most of the night and reaches a stationary point on 7 April, to the north and west of Regulus. Although Mars is fading rapidly as its distance increases (magnitude –0.3 to +0.4) it is much brighter than Regulus (magnitude +1.3). The direct motion of Mars after the stationary point takes it back towards Regulus, and Mars passes less than 2 degrees north of this star on 29 April. See diagram in January notes.

Jupiter is also in Leo, not far from Regulus and Mars. The retrograde motion of Jupiter carries it towards Regulus, but it does not quite reach the star, and it comes to a stationary point about a degree east of Regulus on 26 April. Magnitude –2.0 to –1.8. After this date Jupiter moves direct once more, but its motion at first is quite slow and it is soon overtaken by Mars (see May notes). An unusual event on 9 April, when Jupiter may be seen without any of its attendant satellites, is described in the note below.

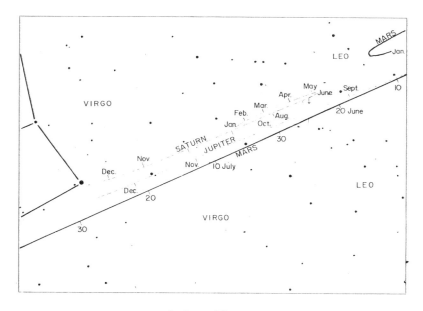

Jupiter and Saturn

Saturn is to be found to the east of Jupiter and Mars and therefore sets later. This will be more noticeable in the southern hemisphere where the ecliptic slopes down steeply to the western horizon in the early morning hours of April. Saturn is still moving retrograde, and passes back into Leo at the beginning of the month. Magnitude +0.9 to +1.0.

Pluto is at opposition on 10 April on the borders of Virgo and Bootes (see diagram in June notes). The very eccentric orbit of this planet has now brought it nearer to the Sun than Neptune, and as a result it is also nearer to the Earth at opposition (see note on Neptune on page 100). The distance of Pluto from the Earth at this opposition is 2714 million miles (4368 million km), but since the orbit is inclined at an angle of 17 degrees, the planet is more than 800 million miles north of the ecliptic. The magnitude of Pluto is $+14$. Photographs taken in 1978 suggest that Pluto has a satellite at a distance of about 20,000 km with a period of 6.39 days. These very difficult observations await confirmation.

THE SATELLITES OF JUPITER

The four great satellites of Jupiter are so big and bright that any small telescope – even an opera glass – will show them. Their ever-changing positions, their disappearance and re-appearance from eclipse or occultation, provide a constant source of interest. If any of them are missing, they are either eclipsed or occulted by Jupiter, or are in transit across the face of the planet. On certain very rare occasions it is possible for all four of these moons to be hidden from view in one or other of these ways, but they cannot all undergo the same phenomenon at the same time. One of these rare occasions occurs on 9 April this year. From $13^h 13^m$ G.M.T. satellite I (Io) will be in transit across the face of Jupiter, satellites II (Europa) and III (Ganymede) will be occulted, and IV (Callisto) will be eclipsed. At $14^h 15^m$ Ganymede reappears from occultation, and for 19 minutes this will be the only satellite visible. Then at $14^h 34^m$ Ganymede will be eclipsed by the shadow of Jupiter, an the planet will remain without visible satellites until $15^h 28^m$, when the transit of Io ends. Thus Jupiter will apparently be without its four big moons for two periods of 62 and 54 minutes. The event is not visible in the British Isles or in America, but observers in Australia and New Zealand should be able to see at least part of this succession of events.

Predictions of this unusual phenomenon, and of many

others of the same kind, were made seventy years ago by an amateur astronomer named Enzo Mora. It would be interesting to know more of this man. His published work, written in French from an address in Italy, appeared in a well-known German astronomical journal. His times were given to a tenth of an hour, but even this accuracy is quite remarkable, and we have no means of knowing how he did the calculations. The tables which we use today had not then been published, and those that were available had been shown to contain considerable errors.

THE SOLITARY ONE

Hydra (the Watersnake, sometimes identified with the serpent killed by Hercules) is the largest constellation in the sky, but it contains only one bright star. This is Alphard (Alpha Hydræ), of magnitude 1.98. It is known as 'the Solitary One' because it lies in a very barren area. It is therefore easy to identify; a line from Castor passed through Pollux and extended for some distance will indicate its position.

Alphard is of spectral type K2, and is therefore obviously orange. The distance is 94 light-years, and the absolute magnitude -0.3, so that Alphard is more than 100 times as luminous as the Sun. Since the declination is only 8½ degrees south of the celestial equator, Alphard is visible from every inhabited continent.

One point of interest is that Sir John Herschel regarded Alphard as decidedly variable. In his diary, compiled during his stay at the Cape of Good Hope during the 1830s (when he laid the true foundations of southern-hemisphere astronomy) he recorded various observations. Thus on his homeward voyage, on 5 May 1838, he wrote that 'Alpha Hydræ is visibly less than last night – it is now fainter than or at the utmost barely equal to Gamma Leonis.' On 7 May he recorded that Alpha Hydræ was well below Gamma Leonis in magnitude: 'Though low, yet it is now decidedly an insignificant star. It is very obviously much diminished since Saturday night ... it leaves no doubt in my mind of the minimum being nearly attained.' By 12 May Herschel made Alpha Hydræ comparable with Castor: 'It would appear that Alpha Hydræ increases more slowly from the minimum than it diminishes to it.' The variability has never been confirmed, and modern catalogues give the magnitude as virtually constant. However, in view of Herschel's comments, it is worth keeping a watch on the star. The main difficulty is in finding a comparable star, at a comparable altitude, to use as a standard.

May

New Moon: 14 May   *Full Moon: 29 May*

Mercury is in superior conjunction on 13 May and will not be seen until the end of the month, when it moves out to eastern elongation. The planet may be glimpsed as an evening star to the north of west after sunset at the end of May. It is quite bright but should not be confused with Venus, which is in the same part of the sky but very much brighter and farther east.

Venus reaches its greatest brilliancy on 9 May (magnitude –4.2), but now sets earlier each evening in the north-west. By the end of May it sets less than an hour after sunset in both northern and southern latitudes.

Mars is an evening star moving direct in Leo and passing less than a degree north of Jupiter on 4 May at 06^h. This is the third of the conjunctions of these two planets, and both will be seen near the star Regulus. Although Mars has lost much of its brightness (magnitude +0.4 to +0.8) it is brighter than Regulus, and easily recognized. (See diagram in January notes). In southern latitudes Mars sets about midnight, but remains an hour or more longer in northern skies.

Jupiter is close to Mars but very much brighter (magnitude –1.8 to –1.6). Having just passed a stationary point, the direct motion of Jupiter is slow – only a tenth of the apparent motion of Mars. The conjunction of Mars with Jupiter on 4 May is closer than the two previous conjunctions, and although Mars has lost much of its brightness the two planets will give added interest to the familiar Sickle of Leo.

Saturn is moving slowly retrograde and reaches a stationary point on 23 May. It is then just inside the constellation Leo, and south of the well-known figure of the Lion. By the end of the month Saturn sets shortly after midnight (magnitude +1.0 to +1.2).

Uranus is at opposition on 14 May and is then in Libra and just visible to the naked eye (magnitude +5.8). The distance of Uranus at opposition is 1649 million miles (2654 million km) and its position is shown in the diagram in the September notes.

URANUS WITH THE NAKED EYE

Uranus, at opposition this month, is above the sixth magnitude, and is therefore visible with the naked eye on a dark, clear night. However, it is by no means an easy object, and it is not in the least surprising that it remained unknown until William Herschel discovered it in 1781. Herschel did not immediately recognize it as a planet; his report to the Royal Society was headed 'An Account of a Comet', and only when the orbit had been computed was its planetary nature recognized. The first to come to this conclusion seem to have been the Finnish mathematician Anders Lexell, and (independently) the French nobleman J. de Saron, who was later guillotined during the Revolution.

Yet Herschel was not the first to see Uranus. Altogether, twenty-two pre-discovery observations have been listed. On 23 December 1690 John Flamsteed, the first Astronomer Royal, recorded it as a star, and even gave it a number (34 Tauri). Flamsteed saw the planet six times in all; James Bradley twice; T. Mayer once, and P. Le Monnier a dozen times. Had Le Monnier been blessed with an orderly mind he could hardly have failed to identify Uranus, but failed to compare his observations, some of which were later found scrawled on the back of an old paper bag!

Uranus and Neptune are often classed as twins, but there are important differences between them (see the article in the 1979

Yearbook by Garry Hunt). In particular, it is likely that Neptune has an internal heat-source, while Uranus has not. The ring of Uranus has been fully confirmed; it is quite possible that Neptune has a ring of the same kind, but Neptune is, of course, much further away, and since it is well below the seventh magnitude it is far below naked-eye visibility.

C.A.F. PETERS

May 8 is the Centenary of the death of Christian August Friedrich Peters, who was born at Hamburg in 1806, studied at Königsberg under Bessel, and then became Director of the Altona Observatory before moving to Kiel in 1878. He was concerned mainly with stellar parallaxes, and confirmed Bessel's work upon the then-unknown companion of Sirius.

GAMMA VIRGINIS

During May evenings the constellation of Virgo is prominent both from Europe and from southern-hemisphere countries such as Australia. The leading star is, of course, Spica. There are two more stars above the third magnitude: Gamma (Arich), 2.76, and Epsilon (Vindemiatrix) 2.86. Of these, Arich is of special interest.

It is a binary, and is only 32 light-years away, so that it is one of our closer stellar neighbours. The components are virtually equal at magnitude 3.6 each, and each is of spectral type FO. Because Arich lies almost at the 'standard distance' of 10 parsecs (32.6 light-years), its apparent magnitude is practically equal to its absolute magnitude.

Several decades ago, Arich was one of the easiest double stars in the sky, and any small telescope would separate it. Today it is not so spectacular. The revolution period is 180 years, and the separation is becoming steadily less, so that by the end of the century the appearance will be that of a single star except when large telescopes are used.

This does not mean that the components are really closing up. Everything depends upon the angle from which we view them, and after minimum separation as seen from Earth the pair will seem to move apart again.

Virgo is crossed by the celestial equator, and is therefore visible from every inhabited continent. Of its leading stars, both Zeta (magnitude 3.4) and Eta (magnitude 4.0) are almost on the equator – only about half a degree south in each case. Apart from Hydra, Virgo is the largest constellation in the entire sky, and covers 1294 square degrees. It adjoins Leo, which at the moment contains three bright planets – Mars, Jupiter, and Saturn, so that the general aspect of this part of the sky is somewhat unfamiliar!

June

New Moon: 12 June   *Full Moon:* 28 June

Solstice: 21 June

Mercury is at greatest eastern elongation (24 degrees) on 14 June and is an evening star to be found in the north-west throughout the month. Observation may be difficult in the bright sky after sunset in the northern hemisphere, but in southern latitudes it should be easier to find, well above the horizon and brighter at the beginning of June.

Venus is in inferior conjunction on 15 June, and then becomes a morning star. By the end of the month it rises north of east about an hour before sunrise.

Mars is an evening star moving direct in Leo, but at the end of June it passes into Virgo. By that time it sets about an hour before midnight, and as it is moving rapidly south, observers in the southern hemisphere have the advantage of seeing the planet in a dark sky for the next few months. Mars passes nearly two degrees south of Saturn on 25 June (see diagram in the April notes) and it will be seen that Mars is then slightly brighter than Saturn (magnitudes: Mars +1.2, Saturn +1.3).

Jupiter is still a bright evening star (magnitude –1.6 to –1.4). moving direct in Leo. In the northern hemisphere Jupiter sets before midnight, and the planet is nearing the end of this period of useful observing. Southern observers have the advantage of seeing Jupiter in a dark sky, although it sets earlier in the late evening.

Saturn is also in Leo, but farther east than Jupiter, and therefore setting later. The planet is moving slowly towards the constellation Virgo, and it is rapidly overtaken by Mars, which passes south of Saturn on 25 June. (Magnitudes: Mars +1.2, Saturn +1.3)

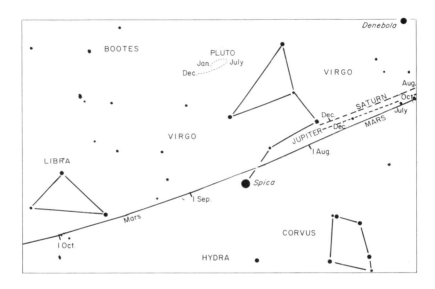

Mars and Pluto

Neptune is at opposition on 12 June, and is to be found in Ophiuchus, which is not usually counted as one of the Zodiacal constellations, although it actually straddles twenty degrees of the ecliptic. The distance of Neptune at this opposition is 2720 million miles (4378 million km), which is greater than the opposition distance of Pluto (see April notes). The distances of the two planets from the Sun at this time are: Neptune 2816 million miles, Pluto 2803 million miles. The orbit of Neptune is almost circular and its distance from the Sun will change very little, but the distance of Pluto will continue to decrease until it reaches perihelion in 1989 at a distance of 2760 million miles. The path of Neptune during the year is shown in the diagram in the September notes, but the planet is not visible to the naked eye, since its magnitude is +7.7.

THE INNERMOST PLANET

Mercury has been known since prehistoric times. The earliest observation of it of which we have any definite record was made on 15 November 265 B.C, when (according to Ptolemy) the planet was close to the stars Delta and Beta Scorpii, but no doubt it was known long before that. It is relatively elusive only because it always remains in the same part of the sky as the Sun, and is therefore seen against a light background. Actually it is bright: the maximum magnitude is –1.9, more brilliant than any star – even Sirius.

Telescopically, very little can be seen on the surface, though the phases are easy to detect (they were first noted by Hevelius, in the seventeenth century). William Herschel tried to distinguish markings, but with no success, and it is likely that the features recorded by his contemporary Johann Schröter were spurious. The first attempt to compile a map was made by G..V. Schiaparelli, between 1881 and 1889. Schiaparelli drew various dark features, and even named them; he believed the rotation period to be equal to the revolution period (88 Earth-days), in which case the same hemisphere would have been turned permanently sunward. This would result in a region of permanent 'day' and an opposite region of everlasting 'night', with an intervening 'twilight zone', over which the Sun would rise and set, always keeping close to the horizon. This twilight zone would have been caused by effects similar to the librations of the Moon, since the orbit of Mercury is more eccentric than that of any other planet except Pluto.

The next serious observer of Mercury was E.M. Antoniadi, who had the advantage of using the fine 33-inch refractor at the Observatory of Meudon, near Paris. In 1934 Antoniadi published a map, which was decidedly different from Schiaparelli's, but which was accepted as the standard until the flight of Mariner 10 in 1974. Antoniadi agreed that the rotation must be synchronous, and he also believed that the planet was surrounded by an atmosphere dense enough to hold material in suspension*.

In 1962, W. E. Howard and his colleagues at Michigan measured long-wavelength radiations from Mercury, and found that the dark side was much warmer than it would have been had it never received any sunlight. The rotation period is now known to be 58.6 Earth-days, or two-thirds of a Mercurian 'year'; when Mercury is best placed for observation from Earth, the same face is always presented to us. However, the high-quality pictures obtained from Mariner 10 during its three active passes (March and September 1974, and March 1975) showed that even Antoniadi's map bore little resemblance to the truth. Moreover, the atmosphere is negligible.

The Mariner 10 pictures covered wide regions of the planet, and craters were detected, as well as mountains, valleys, scarps, ridges, and basins. The most imposing basin, Caloris Planitia, is bounded by a ring of smooth mountain blocks. Unfortunately only about half of it has been studied; the remainder was in darkness each time Mariner 10 flew past Mercury.

In every way Mercury is as hostile as it could possibly be. No manned expeditions there can be anticipated, at least in the foreseeable future, but there can be little doubt that more automatic probes will be sent to it before many years have passed. Over half the total surface remains to be mapped, though there is no reason to expect that the regions still unknown will be markedly different from those surveyed from Mariner 10.

Mariner 10 itself is still in orbit round the Sun, and presumably still makes regular close approaches to Mercury, but it is now 'dead', and there is no hope of re-contacting it in the future.

OPHIUCHUS AND THE ZODIAC

Ophiuchus, the Serpent-bearer, is not counted as a Zodiacal constellation even though it intrudes into the Zodiacal band

---

* For some reason, Antoniadi's book *La Planète Mercure* remained untranslated until 1975, when I produced an English version (*The Planet Mercury:* David & Charles Ltd). It is, of course, of historical interest only, but it was at least a noble attempt!

for some distance. Astrologers, needless to say, take no notice of it – and in addition, precession means that the Zodiacal 'signs' are now out of step with the constellations: the First Point of Aries has shifted into Pisces. Astrology, once regarded as a true science, has long since been discredited, and must be regarded as a subject suitable only for the credulous, even though there are many people who continue tỏ take it seriously.

July

New Moon: 12 July   *Full Moon:* 27 July

Earth is at aphelion (farthest from the Sun) on 5 July at a distance of 94.5 million miles (152.1 million km).

Mercury is in inferior conjunction on 11 July, but by the end of the month it may be found as a morning star rising in the north-east. It may be seen to better advantage in the southern hemisphere, and will be even brighter in the first two weeks of August.

Venus reaches its greatest brilliancy (magnitude −4.2) as a morning star on 22 July. It moves out rapidly from the Sun, and by the end of the month it rises nearly three hours before sunrise. In southern latitudes it may be seen in the dark sky before dawn.

Mars is an evening star in Virgo, but sets in the late evening. By the end of the month Mars will have faded to magnitude +1.4, but is then south of the equator, and observers in southern latitudes will be able to find the planet in a dark sky. The apparent path of Mars at this time is shown in the diagram in the April notes.

Jupiter is still in Leo, but in the northern hemisphere it sets about an hour after sunset and will be a difficult object in the twilight (magnitude −1.4 to −1.3). In southern latitudes the planet sets in mid-evening, and even at the end of July it may be seen for an hour after the end of twilight.

Saturn moves from Leo into Virgo in early July, and then sets in the west in the late evening. Saturn is now only at magnitude +1.4, but on 23 July, the plane of the rings passes for the third time through the Earth. From then onwards both Sun and Earth will be on the north side of the rings, which will once more become visible, and this will add considerably to the brightness of the planet. By the end of the year the rings will have opened to an angle of rather more than 7 degrees.

A penumbral eclipse of the Moon on 27 July will be of little interest, only 28 per cent of the Moon's disk being immersed in the Earth's shadow, See page 128.

THE SCORPION

One of the most magnificent constellations in the sky is Scorpio (or Scorpius), the Scorpion. From Britain and the northern United States it is always inconveniently low down, but from countries such as Australia and South Africa it can be seen to advantage, and is near the zenith during evenings in July.

Scorpio is one of the few constellations to bear at least a vague resemblance to the object it is meant to represent! It consists of a curved line of bright stars, of which the leader is Antares, the so-called 'Rival of Mars'. Antares is a red supergiant, over 500 light-years away, and almost 10,000 times as luminous as the Sun. The colour is unmistakable; in fact Antares is the reddest of all the really bright stars. It has a 6.5–magnitude companion at a distance of 2.9 seconds of arc. The companion looks greenish (due partly to contrast), and has been found to be a radio source.

Antares is flanked to either side by a fainter star (Sigma Scorpio to the north, Tau Scorpio to the south). At the southern end of the constellation is the magnificent 'sting', including Lambda Scorpii (Shaula) which is of magnitude 1.6. In the same area is Iota[1] Scorpii, which is only of the third magnitude, but which is extremely luminous; it may be the equal of 60,000 Suns, and is well over 3000 light-years away, so that if it were as close to us as Antares it would be much more imposing.

The Milky Way flows through Scorpio, and the whole area is extremely rich. There are many star-clusters, of which Messier 6 and Messier 7 are visible with the naked eye.

There is a quaint old legend associated with Scorpio. Mythologically it is said to represent the creature which killed the great hunter, Orion, by stinging him in the heel – thus disproving Orion's boast that he could dispose of any living beast. Subsequently both Orion and Scorpio were placed in the sky, but on opposite sides of it, so that there could be no fear of any further unpleasantness!

HR DELPHINI AND V.1500 CYGNI

In 1967 the famous English amateur astronomer George Alcock discovered a nova in the constellation of Delphinus, the Dolphin (not far from Aquila). It was then easily visible with the naked eye, and at maximum reached magnitude 3.7. Unlike most novæ, it was slow to decline, and remained a naked-eye object for many months. After oscillating around the fourth magnitude it began a steady fade, but this too was gradual, and in 1979 it was still above the twelfth magnitude.

Nova Delphini (now officially listed as HR Delphini) is certainly an exceptional object. Its pre-outburst magnitude was 12, and it may not decline much below this value, in which case it will remain visible with modest telescopes. The nova which flared out in Cygnus in 1975 (V.1500 Cygni) was very different. It brightened up from below the 18th magnitude to above the 2nd with amazing rapidity, and at maximum it reached magnitude 1.8, so that it was more brilliant than any star in Cygnus apart from Deneb. However, its glory was short-lived. Within a week it had fallen below naked-eye visibility, and it is now very faint.

Few novæ of modern times have reached the first magnitude. The list since 1900 is as follows:

1901 GK Persei;	maximum magnitude	0.0
1918 V 603 Aquilæ:	"	−1.1
1925 RR Pictoris:	"	1.1
1934 DQ Herculis:	"	1.2
1924 CP Puppis:	"	0.4

We cannot tell when another bright nova will appear. Neither can we tell when another supernova will flare out in our Galaxy; during the past thousand years only four have been certainly identified – the stars of 1006 (Lupus), 1054 (Taurus: the supernova leaving the Crab Nebula as its end product), 1572 (Tycho's Star in Cassiopeia), and 1604 (Kepler's Star in Ophiuchus). The Lupus star was undoubtedly the most brilliant, and according to R. Stephenson of Newcastle University it reached magnitude −9.5!

August

New Moon: 10 August   *Full Moon:* 26 August

Mercury is at greatest western elongation (19 degrees) on 1 August, and is a morning star to be seen in the north-east before sunrise. The planet is brightest after the date of elongation, but in the northern hemisphere conditions will probably be better at the western elongation in November. Mercury is in superior conjunction on 26 August.

Venus is at greatest western elongation (46 degrees) on 24 August and loses a little of its brilliance as a morning star (magnitude –4.2 to –3.9). In northern latitudes Venus rises about an hour after midnight in a dark sky, and will be well displayed for the rest of the year. The planet is still well north of the equator, and this is a disadvantage to southern observers. At 35 degrees south, Venus rises about an hour before dawn.

Mars is an evening star in Virgo, but in the northern hemisphere it will be a difficult object, setting in the twilight about an hour after sunset, and continuing in this way for the rest of the year. During the month the magnitude is reduced to +1.5, only a tenth of the brightness at opposition. Mars is now south of the equator, and in southern latitudes the planet sets in the late evening in a dark sky. Mars passes 2 degrees north of Spica on 18 August, and this gives an opportunity to compare the magnitudes of both objects (Mars +1.4, Spica +1.2).

Jupiter is in Leo, near the border with Virgo, but in the northern hemisphere it sets shortly after the Sun and is not likely to be seen. In southern latitudes the Sun sets much earlier, and the planet may be seen in the west after sunset in the early days of August. The path of Jupiter shown in the January diagram is continued on the same scale on the diagram in the April notes.

Saturn is moving direct in Virgo, and on 21 August will be about a degree north of the fourth magnitude star Beta Virginis – very much as it was at the beginning of February (see diagram in the April notes). Saturn remains in Virgo until the end of 1983. Although it is approaching conjunction with the Sun, and therefore moving away from the Earth, Saturn is actually a little brighter (magnitude $+1.4$ to $+1.3$) because the rings are opening and adding their contribution to the total light from the planet. In northern latitudes the planet will be lost in the twilight of the western sky, but southern observers have a better chance to see Saturn in the first half of the month, setting in the west in a dark sky.

An annular eclipse of the Sun on 10 August will be visible in Africa and India. Some details are given on page 127.

A penumbral eclipse of the Moon on 26 August will be visible in Europe and the eastern parts of North America. At maximum eclipse 73 per cent of the Moon's disk will be covered by the penumbral shadow. See page 128.

ANNULAR ECLIPSES OF THE SUN

Because the Moon's distance from the Earth is not constant, its apparent diameter varies. When near apogee it looks smaller than the Sun, and when the three bodies are in exact alignment the Moon's dark disk is surrounded by a ring of the Sun's surface: hence the term annular eclipse (Latin *annulus,* a ring). The eclipse visible this month is annular.

Future annular eclipses (1980-1990) are as follows:

1981 February 4.(Pacific, South Australia, New Zealand.)
1983 December 4.(Atlantic, Equatorial Africa.)
1984 May 30.   (Pacific, Mexico, USA, Atlantic, North Africa.
1987 September 23.(Russia, China, Pacific.)
1990 January 26.(Antarctica.)

The photographs given here were taken at the annular eclipse of 29 April 1976, as seen from the Greek island of Thera.

Annular eclipses are interesting to watch, but it cannot be said that they are of real astronomical importance. Neither the corona nor the prominences can be seen, and the sky does not become really dark, though there is a noticeable reduction in sunlight.

THE SUMMER TRIANGLE

To European and North American observers, the evening skies during summer are dominated by Vega, Altair, and Deneb, which make up a large triangle. From Britain, Vega is near the zenith, and cannot be mistaken, both because of its brilliance and because of its bluish colour. The nickname of the 'Summer Triangle' is quite unofficial, but seems to have come into general use. Obviously it does not apply in the southern hemisphere. From South Africa (where it is certainly not summer during August), Altair can attain a respectable altitude, but both Vega and Deneb remain low down, so that they can never be seen to advantage.

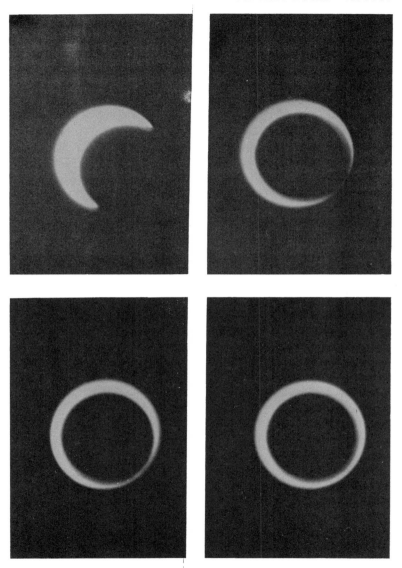

The annular eclipse of 29 April 1976, photographed by Patrick Moore

September

New Moon: 9 September   *Full Moon:* 24 September

Equinox: 22 September

Mercury moves out from conjunction to become an evening star, but in the northern hemisphere it sets shortly after the Sun and is not likely to be seen. In southern latitudes the planet is well placed from mid-September onwards, well above the horizon in the west just after sunset.

Venus now begins to move south with the Sun, but remains a brilliant object in the early morning sky. In the northern hemisphere Venus rises to the north of east about four hours before sunrise, but in southern stations the planet will not be seen before dawn.

Mars passes into Libra in early September, but in the latitude of the British Isles it sets in the twilight little more than an hour after sunset. In the southern hemisphere Mars sets in the late evening, and although it is now only at magnitude +1.5, it should not be difficult to find. The diagram opposite shows the apparent path of the planet up to late November.

Jupiter is in conjunction with the Sun on 13 September, and after this becomes a morning star. It will then be in Virgo, and by the end of the month rises about an hour before sunrise in the northern hemisphere.

Saturn is also in conjunction with the Sun on 23 September and will not be visible during the month.

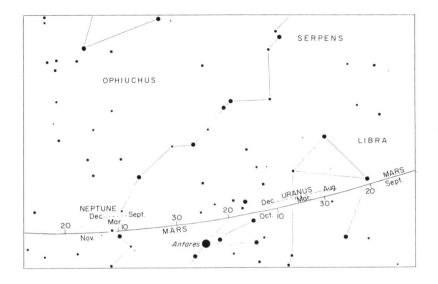

Mars, with Uranus and Neptune

THE DATE OF THE EQUINOX

The addition of the extra day in a leap year causes the dates of the equinoxes in March and September to fall back by one day. These two dates in 1980 are the earliest that have occurred in the present century, but they are by no means the earliest possible. The date of the equinox is not fixed, but is defined as the time when the Sun reaches the First Point of Aries, when its longitude is 0 degrees (or the First Point of Libra, longitude 180 degrees). The actual length of the tropical year (the time taken for one revolution of the Earth with respect to the equinox) is 365.24219 days, so that the date of the equinox moves forward each year by the odd 0.24219 day, which is 11 minutes short of six hours, but it moves back one day in leap years. The same rule applies to the solstices as well as to the equinoxes, so that all four move forward about six hours in common years, and fall back about eighteen hours in leap years. The precise times vary by some minutes because of perturbations and other factors:

113

	March			*September*		
	d	h	m	d	h	m
1972	20	12	22	22	22	23
1973	20	18	13	23	04	21
1974	21	00	07	23	09	59
1975	21	05	57	23	15	55
1976	20	11	50	22	21	48
1977	20	17	43	23	03	30
1978	20	23	34	23	09	26
1979	21	05	22	23	15	17
1980	20	11	10	22	21	09
1981	20	17	03	23	03	05

If we had a leap year every fourth year, as in the old Julian calendar, the average length of the year would be 365.25 days. This is 0.00781 day longer than the true length of the year, and in 400 years the total error would be 3.124 days. To overcome this fault the Gregorian calendar omits three leap years in the 400-year cycle, namely the centennial years which do not divide by 400 (i.e., 1900, 2100, 2200, etc.). The error is then only 0.124 day, or about three hours. It will be more than 3000 years before the error amounts to one day, but this is a problem for future astronomers to solve.

The dates of the equinoxes move slowly backwards during the 400-year cycle of the Gregorian calendar, and there is a range of rather more than two days in the possible dates of each equinox or solstice. The latest dates in the present cycle occurred in 1903 (March 21^d 19^h and September 24^d 06^h). The year 2000 is a leap year, so that the dates will continue to move back until the year 2096, when the equinoxes will fall on March 19^d 14^h and September 21^d 23^h, and this is the earliest that they can be in the present cycle. The year 2100 is not a leap year, and the dates will move forward by one day, as they will also in 2200 and 2300.

BLUE MOONS

Everyone has heard the expression 'Once in a blue moon', but it is not always realized that blue moons can occur! One of the most famous cases on record occurred thirty years ago: on

26 September 1950. For some hours the Moon looked genuinely blue, and the appearance was unusual enough to cause widespread comment. There was no mystery about it. The phenomenon was due to material sent into the Earth's upper atmosphere by vast forest fires raging in Canada. Previously there had been a blue moon on 27 August 1883 (due to the material sent up from the Krakatoa outburst), and green moons were seen briefly from Sweden on two occasions during 1884.

H. C. SCHUMACHER

September 3 1780 was the birth-date of Heinrich Christian Schumacher, who came from Holstein in Germany. He was educated for a legal career, but this did not appeal to him, and he turned to astronomy. He became Director of the Observatory of Mannheim, and later held similar positions first at Copenhagen and then at Altona. He carried out useful work, but is now remembered chiefly as the founder of the astronomical periodical *Astronomische Nachrichten*. Schumacher died on 28 December 1850.

October

New Moon: 9 October   *Full Moon:* 23 October

Summer Time in Great Britain and Northern Ireland ends on 26 October.

Mercury is at greatest eastern elongation (25 degrees) on 11 October and is an evening star. Conditions are not favourable for northern observers, but Mercury continues to be well placed in southern stations, and is to be seen to the south of west until late October.

Venus continues to be a conspicuous object in the morning sky in the northern hemisphere, rising in the east more than three hours before the Sun. Magnitude –3.7 to –3.5. In southern latitudes the planet rises at dawn less than two hours before sunrise. Venus passes half a degree north of Jupiter on 30 October at 20^h. The conjunction should be seen in Australia and New Zealand, but in other countries the two bright planets will be seen together in the morning sky on 31 October.

Mars remains visible in the evening sky, but will be lost in the twilight in the northern hemisphere. In the southern hemisphere Mars sets in mid-evening, but it should be possible to observe the conjunction of Mars with Uranus on the night of 3 October. The actual conjunction occurs at midnight (G.M.T.), and it will only be possible to identify the two planets on the evenings of 3 and 4 October. (See diagram in the September notes). In mid-October Mars passes into Scorpio, crossing the narrow part of this constellation in about 12 days and entering Ophiuchus. Mars passes 4 degrees north of Antares on 24 October at 16^h, and this will afford an opportunity of comparing their colours. Antares is called the 'rival of Mars', and on this occasion is the brighter of the two (magnitudes Antares +1.2, Mars +1.5).

Jupiter is now a morning star moving direct in Virgo. In northern latitudes the planet may be seen in the east at dawn, and by the end of October it will rise in a dark sky. Magnitude –1.2 to –1.3. Jupiter is not well placed for observers in the southern hemisphere, and even at the end of the month is only to be seen in the dawn sky shortly before sunrise. The conjunction with Venus on 30 October is mentioned above.

Saturn is also a morning star in Virgo, and rises a little later than Jupiter. Northern observers again have the advantage of the longer and darker nights, and by the end of October it will be possible to find Saturn (magnitude +1.2) less than five degrees to the east of Jupiter in a dark sky.

THE SQUARE OF PEGASUS

Pegasus, the Flying Horse, is one of the most prominent of the northern constellations. Since it extends to within ten degrees of the celestial equator, it is visible from all inhabited continents. The four leading stars make up the Great Square, though one of them (Alpheratz) has been somewhat illogically transferred to the neighbouring constellation of Andromeda; it used to be known as Delta Pegasi, but has now become Alpha Andromedae. The other members of the Square are Beta (Scheat), Alpha (Markab) and Gamma (Algenib). Scheat is an M-type star, clearly orange in colour, and variable to the extent of about half a magnitude; Alpheratz, Markab, and Algenib are all white.

The apparent magnitudes are 2.1 (Alpheratz), 2.4 to 2.8 (Scheat), 2.5 (Markab) and 2.8 (Algenib). Yet if all were seen from the standard distance of 10 parsecs, the magnitudes would be respectively –0.1, –1.5, –0.1 and –3.4, so that Algenib, which appears the faintest of the four, is actually much the most luminous.

Pegasus contains one other bright star (Epsilon or Enif, magnitude 2.3) and altogether there are nine stars above the fourth magnitude, but the area is not very rich. It is interesting to see how many naked-eye stars are visible within the Square. There are surprisingly few of them.

WILLIAM LASSELL

October 5 is the centenary of the death of William Lassell, one of the most famous of all amateur astronomers. He was born at Bolton, in Lancashire, on 18 June 1799, and became a brewer. His business flourished, and he was able to devote more and more time to his all-absorbing hobby of astronomy. He became an expert mirror-maker, and constructed a fine 24-inch reflecting telescope, equatorially mounted and with a speculum-metal mirror.

In 1846, when the hunt for the new planet Neptune was started, Lassell might well have taken part. By sheer bad luck he was rendered *hors de combat* with a sprained ankle. Had this not been the case, he might well have been the first to identify Neptune; certainly he would have carried out the search with skill and energy. Seventeen days after Galle and D'Arrest had made the discovery, Lassell detected the large satellite of Neptune, Triton. In 1848 he independently discovered Saturn's seventh satellite, Hyperion, and in 1851 he found two new satellites of Uranus, Ariel, and Umbriel. Clearly he was an observer of exceptional ability and keen eyesight.

Lassell became dissatisfied with seeing conditions in the Liverpool area, where he had set up his observatory, and decided to move his large telescope to Malta, where the seeing would be better. He turned his attention to stellar astronomy, and was responsible for discovering six hundred nebulæ. Lassell was also keenly interested in the Moon, and a lunar crater has been named in his honour.

THE START OF THE SPACE AGE

Now that space research has become so much a part of everyday life, it is not always easy to remember that the launching of the first artificial satellite, Sputnik 1, took place only 23 years ago – on 4 October 1957. In October 1959 the Russian probe Lunik 3 went round the Moon, and sent back the first pictures of the far side. The Lunik 3 pictures seem blurred when compared with modern photographs, but they represented a tremendous triumph in those early days, even though some of the details shown were misinterpreted; for instance the 'Soviet Mountains', included the first maps of the far side, do not exist.

November

New Moon: 7 November   *Full Moon:* 22 November

Mercury is in inferior conjunction on 3 November, but moves out to greatest western elongation (20 degrees) on 19 November. The planet is at perihelion on 9 November, so that it is quite bright and should be visible as a morning star for two or three weeks. The diagram shows the changes in altitude and azimuth of Mercury on successive mornings when the Sun is six degrees below the horizon; this is about 40 minutes before

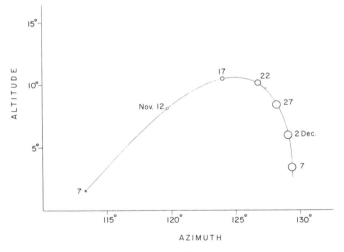

Mercury in November

sunrise in November at latitude 52 degrees north. The changes in brightness are roughly indicated by the size of the circles, and it will be seen that Mercury is brightest *after* the date of western elongation. In the southern hemisphere Mercury is not so well placed, and is likely to be lost in the bright dawn sky.

Venus is still a brilliant morning star, rising in the east about 3 hours before the Sun (northern latitudes). In the southern hemisphere it is no longer to be seen in a dark sky, since it now rises at dawn. Venus passes little more than half a degree south of Saturn on the evening of 3 November, and the two planets will be seen close together in the dawn sky of 4 November. Magnitudes: Venus – 3.5, Saturn + 1.2.

Mars is still an evening star, passing into Sagittarius in mid-November, and remaining in this constellation for the rest of the year. Northern observers are not likely to find the planet, low in the south-west in the twilight, but in the southern hemisphere, Mars remains visible for more than two hours after sunset. Magnitude +1.4. Mars passes two degrees south of Neptune on 10 November (see diagram in the September notes).

Jupiter is a morning star in Virgo, and its direct motion carries it across the equator during the month. In southern latitudes the planet rises shortly before dawn, but in the north Jupiter is a brilliant object in the east in the early morning hours. (Magitude –1.3 to –1.4).

Saturn is also a morning star in Virgo, not far from Jupiter, and also south of the equator. The conjunction of Saturn with Venus is noted above, and the three planets will make an interesting study in the first week of November, as Venus moves quite rapidly past Jupiter and Saturn. (Magnitudes: Venus –3.5, Jupiter –1.3, Saturn +1.2).

FADING STARS

Eridanus, the River, is an immensely long constellation, extending from near Rigel in Orion down to the far south of the sky. The brightest star, Achernar, has a declination of $-57°$, so that it is never visible from Europe.

One of the most interesting stars in the consellation is Acamar, or Theta Eridani. It is of magnitude 2.9, and is a fine double; the magnitudes of the two components are 3.4 and 4.4 respectively, and the separation is 8.5 seconds of arc. Both members of the pair are of spectral type A2, and the distance from us is 65 light years.

According to Ptolemy and other astronomers of Classical times, Acamar was of the first magnitude. If this is correct, then it has faded markedly. Most authorities consider this to be unlikely, but Acamar is not the only star to have been suspected of decreasing. Another is Castor, in Gemini, which was once ranked as being brighter than its neighbour Pollux, but is now considerably fainter (though it is equally possible that Pollux has brightened up). Yet a third example is Megrez or Delta Ursa Majoris, in the Plough or Dipper, which was given as being equal to the other stars in the main pattern – whereas today it is a magnitude fainter.

Stars of this kind are known as 'secular variables', and several other instances have been reported – notably Denebola or Beta Leonis, formerly given as being of the first magnitude while today it is slightly below the second. Whether any of the cases are genuine is very much a matter for debate, but it may be significant that none of the so-called secular variables seem to fluctuate at the present time, and most are of spectral types not generally associated with variability.

We also have cases of stars which are suspected of being variable now, but without definite proof. One of these is Shedir or Alpha Cassiopeiæ, in the famous W or M pattern. It has long been regarded as variable over a small range (perhaps magnitude 2.1 to 2.5), but in the catalogue drawn up by the Russian astronomer Kukarkin it is listed as being constant. Gamma Cassiopeiæ, also in the W, is definitely variable; its normal magnitude is rather below 2, but in 1936 it suddenly increased to 1.6, subsequently fading down to magnitude 3 before starting a gradual recovery. Gamma Cassiopeiæ is an unstable star with an unusual spectrum, and a new outburst may occur at any time.

J. C. WATSON

22 November is the centenary of the death of James Craig Watson a Canadian astronomer who became celebrated for his asteroid discoveries. He was born in Ontario, but then moved to Michigan, and for a time he worked as a factory hand. Subsequently he graduated from Michigan University, and in 1863 became professor of astronomy at Ann Arbor. In the same year he discovered asteroid No. 79, Eurynome, and between then and 1877 he discovered twenty-one others. In those days, of course, all asteroid discoveries were visual (the first discovery to be made photographically, that of 323 Brucia, was delayed until 1891, the successful astronomer being Max Wolf). The 'photographic hunters' of later years were at an obvious advantage; K. Reinmuth's 246 discoveries and Wolf's 232 make Watson's 22 seem minor – but visual searching is laborious by any standards, and Watson spent countless hours at the telescope. He was appointed Director of the Washburn Observatory in 1879, but died in the following year.

December

New Moon: 7 December   *Full Moon:* 21 December

Solstice: 21 December

Mercury is in superior conjunction on 31 December, and will not be seen after the first week of the month, when it is a bright morning star, low down in the south-east. (See November notes).

Venus is now well south of the equator, and rises in the south-east at dawn. The planet is now moving round the far side of its orbit towards superior conjunction in April next. Venus will be at greatest eastern elongation in November 1981, and this will give observers in southern latitudes a particularly favourable opportunity to see this planet as an evening star.

Mars is still an evening star, but sets less than two hours after the Sun, so that it is unlikely to be seen in the twilight. Magnitude +1.4. There is no opposition of Mars in 1981.

Jupiter rises an hour or more after midnight at the beginning of the month, and will be seen in Virgo, moving slowly eastwards towards the planet Saturn. By the end of the year the two planets, close together, will rise at midnight, and Jupiter will pass Saturn in January. This is the first of three conjunctions of these planets in 1981. Jupiter is growing brighter (magnitude –1.4 to –1.6) as it approaches opposition in March.

Saturn can be found in the east of Jupiter and by the end of the year will be close to the third magnitude star Gamma Virginis (see diagrams in April and June notes). As the rings of Saturn slowly open, the brightness of the planet increases (magnitude +1.1 to +1.0), and it will come to opposition in March, less than a day after the opposition of Jupiter The circumstances then are similar to those which arose with Mars and Jupiter in February last, and there will be three conjunctions of Jupiter and Mars in 1981, the first occurring in early January.

ORION

There can be no doubt that Orion is one of the most splendid constellations in the sky – if not the most splendid of all. It is crossed by the equator, which means that it can be equally seen from either hemisphere of the Earth, and its brilliant stars make up a pattern which cannot be mistaken.

The leading stars are:

Star	Apparent magnitude	Absolute magnitude	Spectrum	Distance l/y
Beta (Rigel)	0.08	–7.1	B8p	900
Alpha (Betelgeux)	variable	–5.6v	M2	520
Gamma (Bellatrix)	1.64	–4.2	B2	470
Epsilon (Alnilam)	1.70	–6.8	B0	1600
Zeta (Alnitak)	1.79	–6.6	09	1600
Kappa (Saiph)	2.06	–6.9	B0	2100
Delta (Mintaka)	2.2v	–6.1	09	1500

The distances and absolute magnitudes of these remote stars are uncertain to some extent (some authorities put Rigel at rather less than 900 light-years), but they are certainly of the right order. All the leaders except Betelgeux are extremely hot, and white or bluish-white, while Betelgeux is an excellent example of a red supergiant. Its magnitude range is probably from 0.1 to 0.9, but these are extremes, and the usual value is from 0.4 to 0.7. There is a very rough period of about five years. Mintaka is an eclipsing binary with a very small range.

Quite apart from these brilliant stars, Orion is rich in interesting objects – notably the Great Nebula, Messier 42, in the Hunter's Sword. The nebula is visible with the naked eye as a misty patch, and a small telescope will show it well, together with the multiple star Theta Orionis – known as the Trapezium for reasons which are obvious to anyone who looks at it.

The Nebula was first reported in 1610 by N. Peiresc. It is over a thousand light-years away, and contains an infra-red source which is known as Becklin's Object; the nature of the source is uncertain, but it may be an extremely luminous star which is hidden from us by the intervening nebular material. We now know that the bright nebula is only part of an immense nebular cloud which extends over much of the constellation.

The beauty of Orion is enhanced by its retinue of brilliant neighbours: Aldebaran in Taurus, Capella in Auriga, Castor and Pollux in Gemini, Procyon in Canis Minor and, of course, Sirius in Canis Major. Sirius shines as much the most brilliant star in the sky (it is well over half a magnitude brighter than its nearest rival, Canopus), and when low down it seems to flash various colours, though in reality it is pure white. Star twinkling is purely an effect of the Earth's atmosphere, and has nothing directly to do with the stars themselves; but it is particularly obvious with Sirius because of the exceptional brilliance of the star.

The end of the year

Our calendar depends entirely upon the rotation period and the revolution period of the Earth; the fact that we take $365\frac{1}{4}$ days to complete one journey round the Sun, instead of exactly 365, has led to all manner of complications and modifications in the calendar. Other worlds must have different calendars. Mars, for instance, take 687 Earth-days to go once round the Sun, but its rotation period is longer than ours (24h 37m 22s.6), and so the Martian 'year' contains only 669 Martian days or 'sols'. With Venus, the axial rotation period is longer than the revolution period; Jupiter takes 4332.6 Earth-days to complete one circuit of the Sun, but spins on its axis in less than ten hours.

Eclipses in 1980

The circumstances that control any particular eclipse of the Sun or Moon repeat themselves almost exactly after 18 years 11 days, a period known as the Saros. The eclipses of 1980 are thus very similar to those of 1962, the first year of publication of the *Yearbook*. There will be two eclipses, and since this is the least number possible, both will be eclipses of the Sun. Of less importance are three penumbral eclipses of the Moon.

(1) *A total eclipse of the Sun* on 16 February, the line of totality entering the west coast of Africa to cross Zaire, Tanzania and southern Kenya, the Arabian Sea and southern India and ending in southern China. A partial eclipse will be visible in most of Africa and southern Asia. At Nairobi, which is north of the eclipse path, the eclipse begins at $6^h 52^m$ G.M.T., and by $8^h 22^m$ about 94 per cent of the Sun's disk will be obscured by the Moon. The eclipse ends at $9^h 59^m$. At Calcutta the eclipse begins at $9^h 17^m$, reaches a maximum (95 per cent) at $10^h 27^m$, and ends at $11^h 30^m$.

(2) *An annular eclipse of the Sun* on 10 August will be visible only in the Pacific Ocean and in parts of South America. A partial eclipse will be visible in Hawaii, the south-western parts of the United States, the West Indies and most of South America. At Lima, just north of the line of totality, the eclipse begins at $19^h 15^m$ G.M.T., has its maximum (95 per cent) at $20^h 48^m$, and ends at $22^h 06^m$. At Los Angeles only a small (18 per cent) partial eclipse will be seen from $17^h 27^m$ to $19^h 21^m$.

(3) *Penumbral eclipses of the Moon* seldom attract the attention of amateur astronomers, as there is very little loss of light in the penumbral shadow of the Earth, especially in eclipses of small magnitude. Times for the three penumbral eclipses in 1980 are given as a matter of interest:

1 March	begins 18^h43^m	ends 22^h47^m	mag.	68%
27 July	17^h56^m	20^h21^m		28
26 August	1^h41^m	5^h20^m		73

All times are G.M.T.

Occultations in 1980

In the course of its journey round the sky each month, the Moon passes in front of all the stars in its path and the timing of these occultations is useful in fixing the position and motion of the Moon. The Moon's orbit is tilted at more than five degrees to the ecliptic, but it is not fixed in space. It twists steadily westwards at a rate of about twenty degrees a year, a complete revolution taking 18.6 years, during which time all the stars that lie within about six and a half degrees of the ecliptic will be occulted. The occultations of any one star continue month after month until the Moon's path has twisted away from the star but only a few of these occultations will be visible at any one place in hours of darkness.

Only four first-magnitude stars are near enough to the ecliptic to be occulted by the Moon; these are Regulus, Aldebaran, Spica, and Antares. The occultations of Aldebaran, which began in 1978, continue throughout the year, and will be visible in the Northern Hemisphere. A series of occultations of Regulus which began in November 1979, will also occur each month, moving north and becoming visible in northern latitudes in the latter half of the year. Occultations of the planets Venus, Mars, Jupiter, and Saturn are also predicted during the year. Predictions of occultations of this kind have been made on a world-wide basis since 1937 but in recent years occultations of radio sources by the Moon, and of faint stars by satellites or minor planets have aroused a great deal of interest. The exact timing of such occultations (generally using photoelectric methods) gives valuable information about orbits, sizes, atmospheres and sometimes of the presence of satellites. The discovery of the rings of Uranus in March 1977 was the unexpected result of a successful prediction and observation. It is hoped to issue predictions for at least 60 minor planets for 1980, and we already have data for the occultation of a 12th magnitude star by Pluto on 6 April 1980.

Comets in 1980

The appearance of a bright comet is a rare event which can never be predicted in advance, because this class of object travels round the Sun in an enormous orbit with a period which may well be many thousands of years. There are therefore no records of the previous appearances of these bodies, and we are unable to follow their wanderings through space.

Comets of short period, on the other hand, return at regular intervals, and attract a good deal of attention from astronomers. Unfortunately they are all faint objects, and are recovered and followed by photographic methods using large telescopes. Most of these short-period comets travel in orbits of small inclination which reach out to the orbit of Jupiter, and it is this planet which is mainly responsible for the severe perturbations which many of these comets undergo. Unlike the planets, comets may be seen in any part of the sky, but since their distances from the Earth are similar to those of the planets their apparent movements in the sky are also somewhat similar, and some of them may be followed for long periods of time.

The number of comets under observation in any one year is much greater than is generally supposed. The following table gives the numbers of newly discovered comets, of periodic comets recovered as a result of successful predictions, and of comets followed from previous years.

	1975	1976	1977	1978
New discoveries	13	5	8	11
Predicted and recovered	4	6	11	7
Still under observation	14	16	13	17
	31	27	32	35

The new comets of 1978 included four new periodic comets, while a fifth, discovered by Fujikawa in October, proved to be a return of Comet Denning, which had not been seen since its discovery in 1881. Among the comets expected to return to perihelion in 1980 are the following.

Comet Honda-Mrkos-Pajdusaková was discovered in 1948 by three independent observers – hence the triple name. The comet has a small but very eccentric orbit which takes it inside the orbit of Venus at perihelion. The period is 5.3 years and it has returned on five previous occasions, the last being in 1974.

Comet Wirtanen was also discovered in 1948 at Lick Observatory. It has a period of 5.9 years, and was last seen in 1974 when it was badly placed.

Comet Kohoutek is a new short period comet discovered at Hamburg in 1975. This was Dr Kohoutek's fifth cometary discovery, and it proved to have a moderately eccentric orbit with a period of 6.2 years.

Comet Forbes was discovered in 1929 and has a period of 6.4 years. It has not been seen at each return, and has made only five appearances, the last in 1974.

Comet Reinmuth (1) has a much larger orbit which lies entirely outside that of Mars and has a period of 7.6 years. This comet was discovered in 1928 and has made six appearances, the last in 1973.

Comet Brooks (2) presents an interesting example of the effect of perturbations on the orbit of a comet. The original orbit of this comet had a period of 29 years, but a close approach to Jupiter in 1886 reduced this to 6.7 years when it was discovered in 1889. The comet was last seen for a few months in 1973 (its eleventh appearance) and is due to return at the end of 1980.

Comet Tuttle Tuttle was first seen by Méchain in 1790, but it is named after H.P. Tuttle, who rediscovered the comet in 1858, when orbit computing had greatly improved. The comet was shown to have a large and very eccentric orbit with a period of 13.8 years. At perihelion it is near the Earth's orbit, but at aphelion its distance is well beyond the orbit of Saturn. The comet has made nine appearances, and since the orbit has the high inclination of 54 degrees, it shows little change at each revolution.

Comet Encke has the shortest known period (3.3 years) of any comet, and it makes its 52nd recorded perihelion passage at the end of 1980. It was first seen by Méchain in 1785, but it carries the name of the mathematician Encke, who first computed its orbit and predicted its return in 1822. The orbit of Encke's comet is very eccentric but with modern large telescopes it can be detected each year at opposition.

Several other comets can be observed every year, and these have orbits of small or only moderate eccentricity.

Comet Gunn was discovered at Palomar in 1970 and has a period of 6.8 years, the orbit lying entirely between the orbits of Mars and Jupiter.

Comet Schwassmann - Wachmann (1) has a larger orbit of small eccentricity lying between the orbits of Jupiter and Saturn. It has a period of sixteen years, and is remarkable for its sudden outbursts of brightness. Although normally of magnitude +18, it has been known to brighten by more than seven magnitudes.

Meteors in 1980

Meteors ('shooting stars') may be seen on any clear moonless night, but on certain nights of the year their number increases noticeably. This occurs when the Earth chances to intersect a concentration of meteoric dust moving in an orbit around the Sun. If the dust is well spread out in space, the resulting shower of meteors may last for several days. The word 'shower' must not be misinterpreted — only on very rare occasions have the meteors been so numerous as to resemble snowflakes falling.

If the meteor tracks are marked on a star map and traced backwards, a number of them will be found to intersect in a point (or a small area of the sky) which marks the radiant of the shower. This gives the direction from which the meteors have come.

The following table gives some of the more easily observed showers with their radiants; interference by moonlight is shown by the letter M.

Limiting dates	Shower	Maximum	R.A.	Dec.	
Jan. 1-6	Quadrantids	Jan. 4	$15^h 28^m$	$+50°$	M
Mar. 14-18	Corona Australids	Mar. 16	$16^h 20^m$	$-48°$	
April 20-22	Lyrids	April 21	$18^h 08^m$	$+32°$	
May 1-8	Aquarids	May 5	$22^h 24^m$	$+00°$	
June 17-26	Ophiuchids	June 20	$17^h 20^m$	$-20°$	
July 15-Aug. 15	Delta Aquarids	July 25	$22^h 36^m$	$-17°$	M
July 15-Aug. 20	Pisces Australids	July 31	$22^h 40^m$	$-30°$	M
July 15-Aug. 25	Capricornids	Aug. 2	$20^h 36^m$	$-10°$	
July 27-Aug. 17	Perseids	Aug. 12	$3^h 04^m$	$+58°$	
Oct. 15-25	Orionids	Oct. 20	$6^h 24^m$	$+15°$	M
Oct. 26-Nov. 16	Taurids	Nov. 8	$3^h 44^m$	$+14$	
Nov. 15-19	Leonids	Nov. 17	$10^h 08^m$	$+22°$	
Dec. 9-14	Geminids	Dec. 14	$7^h 28^m$	$+32°$	
Dec. 17-24	Ursids	Dec. 22	$14^h 28^m$	$+76°$	M

M=moonlight interferes

Minor Planets in 1980

Although many thousands of minor planets (asteroids) are known to exist, only 2100 of these have well-determined orbits and are listed in the catalogues. Most of these orbits lie entirely between the orbits of Mars and Jupiter. All of these bodies are quite small, and even the largest can only be a few hundred miles in diameter. Thus they are necessarily faint objects, and although a number of them are within the reach of a small telescope few of them ever reach any considerable brightness. Of these the most important are the 'big four', Ceres, Pallas, Juno and Vesta. Vesta can occasionally be seen with the naked eye, and this is most likely to occur at a June opposition, when Vesta is at perihelion. In 1980, however, there is no opposition of Vesta, and the planet is at its brightest (magnitude +6.8) at the end of the year, when it will be in Leo. Ceres will also be at its brightest in December, reaching magnitude +6.6 during its retrograde motion in Gemini near the star Pollux. Juno is at opposition on 13 January in Canis Minor, near Procyon, but its magnitude is then only +7.7. Pallas has an orbit with the high inclination of 35 degrees, and it can sometimes be seen at quite a distance from the ecliptic. In 1980 Pallas is at opposition on 20 October when it will be in Cetus, but its magnitude is only +7.8.

The largest minor planet orbit recorded to date remains that of Chiron, discovered in 1977 and described in last year's *Yearbook*. Chiron has a period of 50 years and an orbit which comes inside that of Saturn at perihelion, and reaches out nearly to the distance of Uranus at aphelion. By contrast we now know of three orbits which are smaller than that of the Earth. These are 1976 AA, now called Aten, 1976 UA, as yet unnamed, and 1978 RA, which has been named Ra-Shalom. These planets, together with a score or more of minor planets with small but eccentric orbits of this kind, have been recognized in the course of a regular search conducted by Eleanor Helin and her associates, using the smaller (18-inch) Schmidt telescope at Palomar.

Some Events in 1981

In 1981 there will be three eclipses, two of the Sun, and one of the Moon.

4-5 February – an annular eclipse of the Sun, visible in Australasia and the west of South America.

17 July – a partial eclipse of the Moon, visible in Europe, Africa and America.

31 July – a total eclipse of the Sun, visible in eastern Europe, most of Asia and the north-west of North America.

THE PLANETS

Mercury may best be seen in northern latitudes at eastern elongation (evening star) on 2 February, and at western elongation (morning star) on 3 November. In the southern hemisphere conditions are very favourable on 16 March (morning star) and 23 September (evening star).

Venus will continue as a morning star until superior conjunction in April. After this it will be an evening star, reaching greatest eastern elongation on 11 November.

Mars is in superior conjunction on 3 April, and there is no opposition of Mars in 1981.

Jupiter is at opposition on 26 March in Virgo. Conjunction is on 14 October.

Saturn is at opposition on 27 March, also in Virgo, and conjunction is on 6 October.

Uranus is at opposition on 19 May in Libra.

Neptune is at opposition on 14 June and is still in Ophiuchus.

Pluto is at opposition on 13 April on the borders of Virgo and Boötes.

PART TWO

Article Section

Is Life on Earth Unique?

IOSIF SHKLOVSKY

In 1959 I published in a popular Soviet youth paper, *Komoso-molskaya Pravda,* an article along the lines of my hypothesis about the artificial origins of Phobas and Deimos, the two satellites of Mars. It made quite a sensation at the time. It was nothing but a practical joke, however, and the quasi-scientific arguments which I marshalled were of the same type! Nevertheless, I had another and more serious purpose: namely, to attract the attention of the public to the potentially limitless possibilities concerning the conquest of space and boundless expansion. This was the dawn of the Space Age. The first artificial earth satellite, Sputnik 1, had been launched by Soviet scientists eighteen months before the publication of my hypothesis.

Later on, I thought a good deal about the exciting problem of the boundless expansion of intelligence into the universe. Such a process would be certain to increase the energy and material potential of a super-civilization beyond all recognition. Moreover, such an expansion would take place in a short period of time (in cosmic terms, of course). For instance, a solar system can be conquered and developed within a few millennia, and a galaxy within several millions of years. I came to the conclusion that *the activity of such a super-civilization could not have escaped the attention of Earth-based astronomers.* As the Russian parable has it, 'you cannot secret an *awl* in a bag' (I am sure that the British, with their rich folk tradition and renowned sense of humour, have a similar saying). If there were super-civilizations in the Galaxy, then our astronomers would see wonders in the skies – but they don't. So far, every attempt to explain visible phenomena by bringing in intelligent activity has been unsuccessful. Nature is governed by its intrinsic laws, and Man is gradually starting to understand them. If there are no visible signs of highly-developed intelligences elsewhere in the universe, we must conclude that such intelligences do not exist.

This, briefly, is the crux of my argument. I have in fact used the old method of logic known as 'negative proof'. Strictly speaking, the obvious fact that there are no super-civilizations in our Galaxy (or, for that matter, in the entire local group of galaxies), does not necessarily mean that various forms of life, including rational beings, are non-existent except on Earth. But even if they do exist, they have not yet developed technology adequate to communicate with us by radio or by any other method.

The first reason for this which comes to mind is, of course, the non-existence of such beings – or their exceptional rarity, virtually equivalent to non-existence. After all, science has not yet explained the origin of life on Earth! There may be other reasons: perhaps, for instance, inevitable self-destruction at a certain stage of development or a loss of ability leading to degeneration. In any case our own virtual loneliness in the universe is beyond doubt. Even the most optimistic and enthusiastic advocates of the existence of extraterrestrial civilizations 'locate' the closest of them, at something like a thousand light-years away. This is what I call virtual loneliness: there are millions of suns like ours within this enormous radius.

The conviction that we are not alone in the universe had become rooted in human thought long before modern ideas about the nature of the universe had been developed. Incidentally, the 'area' of suggested intelligence is shrinking steadily along with the growth of knowledge. Sir William Herschel believed that there was life inside the Sun, and the belief that rational beings existed on the Moon was widespread as lately as the end of the nineteenth century. We know better today.

Strange as it may seem, the conclusion that there is no life elsewhere makes both scientists and non-scientists feel somewhat uneasy. It seems that many people regard us as 'younger brothers', to be guided eventually by those wiser and more advanced than ourselves. Yet we must reject this whole idea. No visiting, super-intelligent beings will arrive to help us sort out our problems on Earth. The responsibility is ours, and ours alone. Since intelligence is unique, we must consider ourselves as pioneers in the universe. We must take care to preserve our environment – and, of course, the whole concept of global war is absolutely inadmissible.

The realization that humanity is a lonely and singular phenomenon in the universe, coupled with an awareness of the vast scale of the cosmos and the frailness and smallness of our beautiful planet Earth, should become a major moral and ethical factor in our thought. The spurt of consciousness which emerged there as a result of a sequence of exceptionally favourable circumstances should not be allowed to die away; it must flare up like a brilliant torch.

Meteor Streams and Rainfall

D. L. McNAUGHTON

The two words 'meteor' (shooting star) and 'Meteorology' (the study of weather and climate) are both derived from the Greek adjective 'meteoros', meaning lofty. Because rain and shooting stars were both to be observed descending from the heavens, many people in centuries past believed that there was an association between them. However, fifty years ago few would have expected that scientists would eventually wonder whether in fact there was not a relationship between meteors and precipitation, but in a manner very different from anything the Ancients could have imagined. To elucidate, we must firstly take a look at the process by which clouds produce rain.

The physics of rain

In order for rain to form, a mechanism is necessary by which the tiny water droplets making up clouds can be persuaded to combine together into drops large enough to fall as precipitation. Without such a mechanism, the cloud droplets would remain in the sky and eventually evaporate.

In tropical or maritime regions it is not uncommon for cloud droplets to grow into raindrops by coalescence. This will occur if a comparatively few droplets are larger than the rest, causing them to drift and therefore to sweep up and absorb the smaller ones. However, over areas of land, particularly in non-tropical latitudes, most rain commences in ice-crystal form. Because the equilibrium vapour pressure over ice is less than that over liquid water, whenever all three of these phases are present vapour will sublime on to the frozen surfaces, while to preserve the balance liquid water will have to evaporate. In other words the ice particles in a cloud will always grow into snow-crystals at the expense of the water drops until they are heavy enough to fall. They will then become even larger by colliding with other crystals or with water droplets, often melting into rain before reaching the ground.

It is not difficult to maintain clean water drops in a super-cooled liquid state at temperatures as low as –15°C or even colder: the smaller the drop-size the easier it is. For this reason, many small or moderate-sized clouds with tops warmer than –15°C consist entirely of liquid droplets which remain too light to precipitate out. Ice will form only in the presence of special micron-sized aerosols, termed ice nuclei. At temperatures colder than about –25°C there are plenty of impurities in the atmosphere capable of acting as ice nuclei, but between 0°C and –15°C there are few. In order to qualify, the surface of an ice nucleus must be compatible with the crystalline structure of ice; however, at colder temperatures the degree of compatibility is less critical because ice forms more readily. This means that clouds penetrating to high, very cold levels in the atmosphere usually produce rain, whereas smaller clouds whose temperatures are warmer than –15°C sometimes produce negligible rainfall because of a shortage of suitable nuclei.

This reasoning has led to attempts to enhance the precipitation from moderate-sized clouds by seeding them with artificial ice nuclei (nearly always silver iodide crystals), in order to make up the deficit occurring in nature. There are grounds for arguing that the seeding not only converts many of the cloud-droplets into potential rain, but it also causes additional latent heat of freezing to be released, thus prolonging the cloud life-time. Sometimes the quantity of extra heat energy produced may be sufficient to increase the buoyancy of the cloud to the stage where it will grow further, reaching colder regions of the atmosphere where the rain-forming process is more efficient.

Unfortunately it is far from easy to establish whether a particular seeded cloud has in fact yielded more rain than if it had been left alone. This is because of the high variability of natural precipitation: a moderately tall cumulus cell will sometimes fail to rain, whereas on other occasions it will produce a heavy downpour. Although there is agreement amongst scientists that seeding can induce changes in certain clouds, there is still debate regarding the extent to which the rainfall can be influenced.

Calendar singularities in precipitation·

There is good evidence, at least in the northern hemisphere, that most natural ice nuclei in clouds consist of soil particles, such as kaolinitic clays. However, E.G. Bowen has suggested that another important source is meteoric dust from outer space. He was led to this hypothesis after noticing a number of calendaricities in rainfall first in Australia, and then in other parts of the world. The Sydney rainfalls between 1859 and 1901 were totalled for each date in the year, and showed two exceptionally large peaks on 12 and 22 January, which departed about four standard deviations from the mean rainfall (Bowen, 1953). He then examined later records at Sydney, between 1902 and 1944, and found that prominent peaks occurred on 12/13 January and 23 January, in other words on almost the same dates as in the completely different earlier period. The same behaviour was later observed by O'Mahony (1962) in four separate 25-year periods at Sydney and two such periods at Rockhampton. O'Mahony (1962) also carried out a detailed analysis of 100 years of Sydney January rainfall. He totalled up the falls recorded on each calendar date excluding first the heaviest fall on each date, then excluding the first and second heaviest, and so on. To his surprise, the residual data continued to display peaks on 12/13 and 23 January even after the ten heaviest falls of every date had been removed.

Bowen extended his investigation to other countries, examining over 100 years of records from Australia and New Zealand, 75 years from the U.S.A. and Japan, 50 years from South Africa and the Netherlands, and 30 years of data for the British Isles (Bowen 1956a and b). All seven countries exhibited rainfall maxima on or near the 12/13 January, and five of them showed a peak near 22 January. A number of other calendaricities also appeared to be too widespread to be attributed to mere coincidence, for example 18 November and 1/2 February. A subsequent paper by Brier (1961) confirmed these and other smaller Bowen singularities in American rainfall during 1952-58, a period which was not included in Bowen's study.

It seemed most unlikely that this world-wide pattern could have resulted from chance occurrences of local heavy rainfall. Bowen therefore argued that only an extra-terrestrial influence could produce nearly simultaneous effects over the entire surface of the Earth. Because the required mechanism would have to possess a yearly periodicity, he put forward an explanation in terms of meteoric particles from cometary orbits acting as ice nuclei in terrestrial clouds. Thus, Bowen associated the rainfall peak near 13 January with the Geminids meteor display of 13 December, the 23 January rain maximum with the Ursids of 22 December, the peak on 1/2 February with the Quadrantids of 3 January, and that of 18 November with the Orionids. The lag of approximately 31 days represented the time taken for the meteoric dust to settle through the Earth's atmosphere to a level where it could nucleate the potential rainclouds.

Supporting evidence for the existence of rainfall singularities came from a study of 22 years' data at six stations in the USSR by Dmitriev and Chili (1958). Of the fifteen rainfall peaks which they found between 1 August and 5 February (i.e. the period covered by Bowen), nine corresponded closely with Bowen's rainfall maxima, while four others showed up in some of the countries investigated by Bowen.

In Rhodesia, an analysis by McNaughton (1970) of the pre-1946 records of 55 stations showed eight rainfall singularities, which were confirmed by hail calendaricities in the completely different period 1957-68. All but the smallest of these singularities coincided with rainfall peaks in other parts of the world. It was possible to demonstrate that the same calendaricities were built up independently in northeast and in southwest Rhodesia from rain falling in different years (McNaughton, 1971). A similar phenomenon had been noted in a comparison of Sydney and Rockhampton rainfalls in Australia by O'Mahony

(1962). This is certainly consistent with the hypothesis of a calendaricity in a secondary rain-forming factor such as atmospheric ice nucleus concentration.

By collating over 300 years of dates when snow first covered the ground at Tokyo, Bowen (1956d) managed to study a period considerably longer than most meteorological records. Calendaricities again appeared on 12 and 23 January, and on other dates when rainfall peaks had occurred (although less consistently) in his earlier investigations.

Long-period cycles in rainfall singularities

The aspect of Bowen's hypothesis which has perhaps received the most criticism is the postulated 30 or 31-day time lag between a visible meteor display and a rainfall anomaly. The meteoric dust particles cover a wide spectrum of sizes; only those with diameters of about eight microns might be expected to settle down into the troposphere in the required time, and even then it would be fair to ask whether atmospheric motion would not cause significant variations in the descent speed. Dr Bowen himself admits that there are gaps in his theory.

On the other hand it would be unfair to omit all mention of a phenomenon which does lend at least some support to the proposed 31-day lag. This is the apparent six- to seven-year periodicity in rainfalls approximately 31 days after a meteor shower exhibiting a similar periodicity. Associated with the remnants of Biela's comet are the Bielids (sometimes also named the Andromedids after their radiant), which produce visible displays in late November and early December. The meteoric particles in the cometary orbit are concentrated in a swarm with an orbital period of about 6.5 years, similar to that possessed by the comet before it became defunct. Thus, the Bielids made brilliant appearances in 1867, 1872, 1879, 1885, 1892, and 1899, when the Earth passed through the denser section of the meteor stream, in other words at intervals of five, seven, six, seven, and seven years. (Since 1899 this shower has not been prominent.)

For the three last days of December (i.e. about a month after the Bielids' display), Bowen (1956a) examined rainfalls from the U.S.A. between 1871 and 1950, and from Australia and New Zealand between 1900 and 1950; in all three countries there was strong evidence for a six-year cycle in heavy rainfall on these dates. In the U.S.A. the correlation between 29 to 31 December rainfalls six years apart was +0.35; the probability of this having occurred purely by chance is as low as one in a thousand. There may have been a phase displacement of about a year between the rainfall peak and that of the meteor display, possibly because the particles affecting the rain were smaller than (and had gradually become displaced from) those giving the visible display. In this connection it is interesting to note that although the meteor displays have almost faded away since 1899, the effect on rainfall appears to have continued at least up until 1948.

The variations in the year when snow first covered the ground at Tokyo on 29, 30 and 31 December also showed evidence of a six or seven-year periodicity which correlated with that of the Bielids (Bowen 1956d).

The Giacobinids (or Draconids) are a northern hemisphere shower associated with the Giacobini-Zinner comet, whose orbital period is 6.6 years. The Giacobinid meteor display usually takes place on or near 9 October; it was particularly prominent in 1913, 1926, 1933, and 1946. The dates of heavy November rain were extracted for 48 U.S.A. stations: it was found that the rainfalls of 8 to 10 November produced sharp peaks in 1913, 1919, 1926, 1932, 1943 and 1947 (Bowen 1956b). In other words the rain appeared to follow the same cycle as the meteor display, but about 31 days later. The rainfalls were also examined on earlier and later dates in November, but did not exhibit this periodicity.

The Perseids are another meteor stream incident on the northern hemisphere, with a period estimated at about 110 years; their activity was at a minimum in 1911. Bowen (1957) examined the records of 48 American stations on the dates 10 to 24 September (i.e. about a month after the Perseids' visible display), and found that the rainfall on these dates had decreased steadily after 1870 until (like the meteors) they reached a minimum in 1911, since when they have increased again.

Measurements of the atmospheric concentration of magnetic spherules

Whipple and Hawkins (1956) questioned Bowen's hypothesis on the grounds that the influx of sporadic meteors detected by radar and photography is of the same order as that of the regular, shower-derived meteors. However, they mentioned the possibility that this might not apply to the particles too small to be measured optically or by radar, but more likely to qualify as ice nuclei in tropospheric clouds. Bowen (1953) supports the idea that the noctilucent clouds of the ionosphere consist of meteoric dust, but the implications of this still need to be investigated more thoroughly.

During selected periods between 1967 and 1971, Rosinski (1970 and 1975) measured the concentration of tiny magnetic spherules in the atmosphere at between four and eleven different latitudes, some north and others south of the equator. On each day at every station about 1000 m³ of air was passed through a polystyrene filter, which was then dissolved in chloroform. The magnetic particles suspended in the solution were separated with a magnetic stirring rod, after which they were sized and counted under a microscope. Their diameters ranged between about 3 and 25 microns.

High particle concentrations appeared on similar dates at widely separated stations, proving that the particles were of extra-terrestrial origin. Sampling took place during October in three different years, all of which showed peaks on or near the seventh of that month, confirming one of Bowen's prominent rainfall singularities (Bowen, 1957). Unfortunately no other month was sampled more than once, but high concentrations of spherules were nevertheless measured on several dates near Bowen anomalies, including 9 to 14 January and 16 to 21 January. However, Rosinski (1970) mentions that Bowen's 30-day time lag is unlikely to apply to all sizes of particles collected on a particular calendaricity; in other words the differently sized spherules were probably derived from separate meteor belts.

Ice nuclei

Various methods have been devised of estimating the number of ice nuclei in the atmosphere. The most common technique makes use of an insulated cold chamber into which a sample of air is drawn. A 'cloud' will form in the chamber as the air cools by convective mixing (or by deliberately reducing the pressure); alternatively steam may be pumped in. The number of ice crystals forming and falling out of this cloud can be estimated by illuminating them in a beam of light, or by allowing them to settle onto a glass tray. Sometimes the tray is filled with cold sugar solution which preserves and enables the ice crystals to grow, so that counting becomes easier. At $-15°$ C the ice nucleus concentration of the atmosphere is often less than one per litre; at $-25°$ C it might be as high as 10 or even 50 per litre.

A programme of January ice nucleus measurements was organized during 1954 to 1958 at stations in Australia, America, and South Africa. The results of these and other observations have been summarized by Bowen (1956c) as well as by Kline and Brier (1958). In every one of the five years, high ice nucleus counts were recorded on or near 13, 22/23 and 31 January, agreeing with the rainfall singularities found by Bowen (1956a) and others. Furthermore, these ice nucleus peaks always appeared in at least two different continents almost simultaneously. Although there were occasions when some of the ice nucleus singularities did not show at certain stations, it is difficult to see how the apparent recurrence tendency cannot be genuine.

It is also relevant to note that many of the counts were made in aircraft at high altitude, which does not in itself prove that the nuclei were of extra-terrestrial origin, but at least it adds to the plausibility of this hypothesis. There is evidence too that the occurrence of cirriform clouds over Australia shows preferred dates on 12 and 22 January and 1 February (Bigg 1957a and b); these clouds form at very high levels in the troposphere and consist entirely of ice crystals.

In months other than January ice nucleus counts have been much less widespread; however, tentative support for Bowen has come from measurements made by Rosinski (1967) between December 1964 and April 1965.

Laboratory experiments by Bigg and Giutronich (1967) indicate that meteoric dust may well be capable of providing ice nuclei. Critical of other experimenters who crushed and therefore destroyed the surface properties of particles being tested, they boiled meteorites at a pressure of only 2 mm of mercury (thereby attempting to simulate conditions in the outer atmosphere), collecting the recondensed material on slides. The particles obtained from a metallic meteorite were of similar size, colour and shape to those sampled by Rosinski (1975), and when tested in a cold chamber at $-14°$ C they gave a count of five ice crystals per thousand spherules. A stony meteorite produced much smaller particles with a lower nucleation rate.

Discussion and Conclusion

More work remains before the connection between meteor dust and rainfall can be regarded as conclusively proved. For example, there is an unexplained discrepancy between the particle concentrations measured by Rosinski (1975), the nucleation rate estimated by Bigg and Giutronich (1967), and actual atmospheric ice concentrations (e.g. Bowen, 1956d). Nevertheless, as yet no other serious proposal has been put forward to account for the rainfall singularities.

Certainly there does seem to be strong evidence that these calendaricities are genuine climatic phenomena, at least during January. This was the month chosen by Bowen for the ice nucleus measurements, and it was the only month considered by O'Mahony (1962) to exhibit statistically significant rainfall peaks in Australia. January is also a month with a well-defined warm temperature singularity near 22 January in the U.S.A. (Wahl, 1953), which may somehow be associated with the rainfall singularity then (Bowen, 1956a).

Bowen suspected that his proposed meteor/rain relationship would be less clear-cut between May and September, because of greater difficulty in distinguishing one meteor shower from another during those months (Bowen, 1953 and 1957). Even in December, one of the months selected by him, the evidence for world-wide singularities is less impressive than in November,

January, and early February. Bowen (1956a) mentions that the Bielids I meteor shower retrogresses in time by one day every five years, so it is pertinent to ask whether the corresponding 'favoured date' 31 days later might not also retrogress, thus questioning the value of analysing the December rainfalls accumulated on fixed calendar dates.

Another question which deserves serious consideration is the possibility of the meteoric source of ice nuclei attaining greater importance in the southern hemisphere, because of the smaller area of land and the consequent dearth of soil-derived nuclei. It is unlikely that the southern continents rely on the northern ones for a large part of their ice nucleus ration, considering the comparatively slow trans-equatorial exchange of air (McNaughton, 1971).

Finally, even though rainfall calendaricities are almost certainly real, caution should be exercised when attempting to use them to predict weather (the same is, of course, true with the recently discovered correlation between moon phase and rainfall). If the historical records of a particular area do not show evidence of a rainfall singularity known to exist elsewhere, then it may well be dangerous to infer that it will show up in that area in the future. Even when the existence of a rainfall anomaly is established using past data, such as 22/23 January at Sydney, it does not manifest itself often (O'Mahony, 1962). To illustrate, rainfall in excess of half an inch was recorded only 13 times on 23 January at Sydney between 1861 and 1960; on 58 occasions no rain fell at all. (Similar figures also apply to the 22 January). This is perfectly reasonable if (as seems likely) the rainfall anomaly is associated with an ice nucleus anomaly, because the nuclei are useless unless the synoptic weather conditions are such that heavy clouds are already present in the sky.

References

Bigg, E.K., 1957a. 'January anomalies in cirriform cloud coverage over Australia'. *J. Meteor.* **14**, 524-526.

—1957b. 'A new technique for counting ice-forming nuclei in aerosols'. *Tellus* **9**, 394-400.

—and J. Giutronich, 1967. 'Ice nucleating properties of meteoritic material'. *J. Atmos. Sci.* **24**, 46-49.

Bowen, E.G., 1953. 'The influence of meteoric dust on rainfall'. *Austral. J. Phys.* **6**, 490-497.

—1956a. 'The relation between rainfall and meteor showers. *J. Meteor.* **13**, 142-151.

—1956b. 'A relation between meteor showers and the rainfall of November and December'. *Tellus* **8**, 394-402.

—1956c. 'January freezing nucleus measurements'. *Austral. J. Phys.* **9**, 552-568.

—1956d. 'A relation between snow cover, cirrus cloud, and freezing nuclei in the atmosphere'. *Austral. J. Phys.* **9**, 545-551.

—1957. 'Relation between meteor showers and the rainfall of August, September and October'. *Austral. J. Phys.* **10**, 412-417.

Brier, G.W., 1961. 'A test of the reality of rainfall singularities'. *J. Meteor.* **18**, 242-246.

Dmitriev, A.A. and A.V. Chili, 1958. 'On meteor streams and precipitation' (in Russian). *Ak. Nauk. Mor. Gidrofiz. Inst., Trudy* **12**, 181-190.

Kline, D.B. and G.W. Brier, 1958. 'A note on freezing nuclei anomalies'. *Mon. Wea. Rev.* **86**, 329-333.

McNaughton, D.L., 1970. 'Calendar singularities of rainfall in Rhodesia'. *Proc. Trans. Rhodesia Sci. Assoc.* **54**, 99-107.

—1971. 'Calendar singularities in Rhodesian precipitation, and the implications'. *J. Appl. Meteor.* **10**, 498-501.

O'Mahony, G., 1962. 'Singularities in daily rainfall'. *Austral. J. Phys.* **15**, 301-326.

Rosinski, J., 1967. 'On the origin of ice nuclei'. *J. Atmos. Terr. Phys.* **29**, 1201-1218.

—1970. 'Extraterrestrial magnetic spherules: their association with meteor showers and rainfall frequency'. *J. Atmos. Terr. Phys.* **32**, 805-827.

—C.T. Nagamoto and M. Bayard, 1975. 'Extraterrestrial particles and precipitation'. *J. Atmos. Terr. Phys.* **37**, 1231-1243.

Wahl, E.W., 1953. 'Singularities and the general circulation'. *J. Meteor.* **10**, 42-45.

Whipple, F.L. and G.S. Hawkins, 1956. 'On meteors and rainfall'. *J. Meteor.* **13**, 236-240.

Observing the Sun

PETER J. GARBETT

I wonder how many times an account written about the Sun has begun, 'the Sun is our nearest star', and how often this has been taken for granted? It is the obvious which is frequently the most important, and this is the case with the Sun. The importance of observing the Sun stems from the simple fact that it is the only star which presents an observable disk. The other stars are so remote relative to the Earth, that out of the 100,000 million or so members of the Galaxy, the Sun is our only immediate hope of learning about stars' surfaces. This is unfortunate, since we cannot be at all sure that the Sun is typical of stars in general, and it would be foolish for us to make that assumption, since wherever we look in the universe we see great diversity.

Quite apart from the cosmological importance of the Sun to the astronomer, the most obvious significance to us, as human beings, is that we owe our existence to the vast quantities of energy it produces, and this alone makes it a more-than-worthwhile field of research. Evidence is also mounting in support of the idea that there is a link between sunspot activity and terrestrial meteorology; and we have known for some time that the Sun is responsible for terrestrial magnetic disturbances and aurorae. We also hear more and more today of the need to find some relatively inexhaustible source of energy when our mineral resources run out, and solar energy may provide some of the answers.

The Sun is a perfectly ordinary star, with a diameter of about 865,000 miles (1,392,000 kilometres) lying on average at a distance of 92,957,000 miles (149,600,000 kilometres) from the

Earth. By volume the Sun would enclose about 1,303,600 planets the size of the Earth, but since its mean density of $1.409g/cm^3$ is substantially less than the Earth's overall density of $5.517g/cm^3$, it is 'only' some 332,946 times heavier. These facts make the Sun approximately average by cosmic standards. There are stars such as Betelgeux in the constellation of Orion, the Hunter, with diameters as large as 225 million miles (360 million kilometres), while there are others, such as Proxima Centauri, with a luminosity only one ten-thousandth that of the Sun. The Sun produces energy by a process known as fusion, as distinct from the method of fission used in the modern nuclear power station. It has been estimated that using this method of producing energy, the Sun will be able to continue shining for another 5,000 million years much as it does today, before becoming a red giant and swallowing up the Earth. Since the Earth is about 4,600 million years old, we have no need yet of packing our bags for another world!

Probably one of the best-known early telescopic observers was the Italian astronomer, mathematician and physicist Galileo Galilei. At the beginning of the seventeenth century, when he began to carry out his first telescopic observations of the heavens, he turned his attention to the Sun. Towards the end of the year 1610 he noted some darker patches on the surface of the Sun, but he did not publish his findings until May 1612, by which time Harriot, Fabricius, and Scheiner had all independently observed and recorded their presence. The darker patches, observed by Galileo and his contemporaries, are known today as sunspots, and we now know that they had been recorded long before the invention of the telescope by the ancient Chinese. Their records show that they almost certainly observed what must have been fine examples of naked-eye sunspots. Modern investigations have told us more about these spots which appear periodically on the surface of the Sun, and today we think of them as centres of violent magnetic disturbances. The temperature of the photosphere, or bright surface of the Sun, has been estimated as being 5,800K*, while that of

*K = Kelvin. Degrees centigrade = K + 273.16.

the umbra, or central dark region of a sunspot, has been shown to be at least 1,000K cooler. This explains why sunspots appear dark, although we must be cautious here, since if a sunspot was allowed to shine on its own it would be brighter than an arc-light! We find that a typical sunspot consists of two main parts; the umbra and penumbra, the latter forming a lighter region about the darker umbra. As expected, the temperature of the penumbra is only slightly less than that of the surrounding photosphere, at 5,500K.

The early observers of the Sun were soon to notice that sunspots frequently occurred in groups, and that they were carried from the eastern edge of the Sun across the disk to the western edge in about 14 days. They also discovered that the life spans of spots varied greatly. Some only lasted for a matter of hours, never developing past the stage of a very small pore, while others developed into much more complex groups, lasting for weeks or even months. Some lasted long enough to be carried round a complete revolution in just under a month, and occasionally became large enough to be visible to the naked-eye. (A sunspot becomes visible to the unaided eye when it attains an area greater than 500 millionths the area of the visible hemisphere.) The actual axial rotation period of the Sun is 25 days at the equator and about 31 days at the poles, showing that the Sun is, of course, a gaseous body. However, since the Earth is itself in orbit around the Sun, a sunspot appears to take 27 days to complete one rotation at the equator.

Associated with sunspots are brighter patches known as faculæ. These are believed to be elevated above the general surface, and are seen at their best around the limb regions of the Sun, where the effect of 'limb darkening' contrasts strongly with their relative brightness. Observation of faculæ over a period of time shows that they are often spread out over a large area of the Sun's surface, appearing before a sunspot group develops, and lasting some time after the sunspot group itself has disappeared. Faculæ occur usually in the sunspot latitudes, that is from about 40° N to about 40° S and it is rare for either a sunspot group or faculæ to be found in a latitude closer than this to the poles. However, at times of very low sunspot activity ('solar minima'), isolated small patches of so called 'polar faculæ' may appear in unusually high latitudes, although these are never associated with sunspots.

After observing the Sun over a long period of time it began to be apparent that the number of sunspots visible on the Sun varied in a periodic manner. The first astronomer to suggest this was Heinrich Schwabe of Dessau, who published his findings in 1851. Although other early observers realized that the number of sunspots varied with time considerably, Schwabe was the first to suggest a periodicity of just over 11 years on average. This 11 year cycle is not always strictly adhered to. For example after the maxima in the year 1787 there was a huge gap of 17 years before the next maxima in 1804, and the maxima of 1830 was followed only 7 years later by another maxima in 1837. Also the height of each maxima is seen to vary greatly. The most active year ever recorded since records began was 1957, where the height of the maxima was three times the height of the 1905 maximum, for instance. This has led some astronomers to believe in a much more long-term cycle, but since our records are only taken over a few centuries, it is difficult to establish the truth of these matters. We do, however, know for certain of a 22-year cycle in the magnetic fields of sunspots. For example with sunspot groups of the bipolar type we find that the leading and following components are of opposite polarity. Thus if a bipolar group in the northern hemisphere had a leader which was a 'south-seeker', the follower would be a 'north-seeker'. In the southern hemisphere this would be reversed, with the leader a 'north-seeker' and the follower a 'south-seeker'. However, at the end of the cycle the situation would become reversed completely, with the leading spot in the northern hemisphere being a 'north-seeker' and so on. After two solar cycles, that is 22 years, the situation will have returned to normal.

In addition to the variations in the number of sunspots visible at any given time in the solar cycle, and to the polarities of spots, it was found that the latitudes in whch spots appeared varied in a regular fashion. At the beginning of a new cycle, that is to say shortly before solar minimum, spots appear in the higher latitudes around ±30° As the cycle progresses, spots appear in increasingly lower latitudes, reaching about ±15° at maxima, and ±8° near minima. Before the low-latitude spots of the old cycle have died out near the equator, spots of a new-cycle have broken out in the high latitudes, producing two zones of activity. (This is known as Spörer's Law.)

We have already looked at what can be seen on the Sun in ordinary white light, but have as yet excluded what can be seen in wavelengths of light beyond the range of the human eye. This is a rather specialized area of work, and I will not dwell on it too long.

Most of us are familiar with photographs of what are known as solar prominences and solar flares. Prominences are great masses of glowing red hydrogen gas, looking like giant flames, which leap off the surface of the Sun, and extend many thousands of miles into the chromosphere (region of the Sun's atmosphere above the photosphere, but lower than the corona). Flares are violent, short-lived outbursts at the solar surface, often associated with active sunspot groups. Neither prominences nor flares are visible under ordinary conditions; however, prominences can be seen during a total solar eclipse, and some observers have claimed to have seen particularly bright flares in white light. Some keen amateurs who are prepared to pay for the pleasure of observing these phenomena, can obtain relatively cheap narrow band filters from the United States for about £100 upwards. These filters have maximum transmission at 6,563 Ångströms (656.3nm), that being the position of the prominent H-α line in the solar spectrum.

In the first half of this article I have dealt with what we already know about the Sun, and I felt that this was necessary before describing what the amateur solar observer can achieve. The obvious advantage the amateur has with the Sun, is that unless the observer comes from the north or south poles, where the Sun will be below the horizon for months, it can be observed during the day-time. This can be a distinct advantage to those who find they need their beauty sleep! However,

BEWARE! The Sun, if observed through a telescope or binoculars, can result in the observer becoming permanently blind. So please heed the warning: **NEVER LOOK AT THE SUN DIRECTLY THROUGH A TELESCOPE OR BINOCULARS**, whether a so called sun-filter is used or not. Filters can and will crack under the heat of the Sun, as a number of people have found to their cost. (The famous observer Galileo may have damaged his sight badly using a telescope, which by modern standards was very small.) The obvious alternative to direct observation through a telescope or pair of binoculars is to project the image of the Sun on to a white screen, where drawings and measurements are possible. If the screen is suitably enclosed inside a box with one side removed the glare can be reduced to a minimum. Such a projection-box is an invaluable aid to the solar observer, and if it is made light enough, then the balance of the telescope need not necessarily be upset too much.

Projection-box on the author's 8¾-inch (22.2cm) Newtonian reflector.

After the amateur has learned roughly what to expect on the Sun, the next thing he or she does is to discover what sort of telescope is suitable for actual observation. Opinions differ greatly on this subject, and the best I can do is to put forward the ideas which I have gained from personal experience. The most fundamental factor that must be appreciated, is that unlike faint galaxies, quasars, and so forth, which require the power of the largest and most expensive telescopes in the world, the Sun provides the astronomer with such a great amount of light, that even the most insignificant and inexpensive telescopes will bring great rewards. A two- or three-inch (5.1cm or 7.6cm) refractor will show a great deal of detail, as will a four-inch (10.2cm) reflector, although most would prefer a refracting telescope in the 4- to 6-inch range (10.2 to 15.2cm). Many amateurs also seem to be under the misapprehension that the Newtonian reflecting telescope is not suitable for solar work. From my own experience I would say that moderate or small reflectors, if used with a little common sense, can be just as effective, although inch for inch the refractor comes out best. There is also the danger with moderate-sized reflectors of getting trouble with over-heating of eyepieces and flats, but this can be avoided by 'stopping-down' the telescope by placing a board with a suitably sized circular hole in the centre over the telescope, effectively reducing the aperture of the instrument. The degree of stopping-down of reflectors will depend on the weather, since on cold winter days, with the Sun low in the sky, the heat will not be sufficient to do any damage; whereas in the middle of summer damage can be done if adequate caution is not taken.

Having selected a telescope which is suitable for observing the Sun, the next step is to build a projection-box. This should be made as light as possible, to avoid counter-balance problems, but strong enough to support a white screen. With a refracting telescope the screen is placed behind the eyepiece, and is basically a continuation of the tube itself; whereas with a Newtonian reflector the projection-box is constructed perpendicular to the line of the tube. If the telescope has an equatorial mounting, and is provided with a drive, a projected image of the Sun can be retained on the screen without making any manual adjustments. The size of image produced in the projection-box on the screen will depend on the eyepiece used (eyepieces with a short focal length will produce a large image), and on the distance between the eyepiece and the screen. I use a magnification of about fifty, with a resultant image of 6 inches (15.2cm) diameter; this I find ideal for drawing purposes. With small telescopes of less than three inches (7.6cm) aperture, the disk size adopted may have to be reduced to 4 inches (10.2cm). Having thus worked out the rough dimensions for the projection-box, the next task is the actual construction. This part I leave completely up to the individual, since there are a whole variety of different methods all with their own relative merits. From my own experience I have found that wood is just as good as any other material, provided weight can be kept down. All that is then needed is a simple rectangular box, with provision at one end for the screen to be held in place. If the observer's interests span a wide area of different aspects of astronomy, the projection-box should be detachable, so that the eyepiece can be got at easily. Allowance should also be made for the fact that the Sun's angular diameter varies between 32' 35" at perihelion, and 31' 31" at aphelion; so that the screen itself, although held firmly, must be able to move backwards and forwards to a small extent.

Having spent time and money on equipment, the next stage is to begin actual observing, and keeping a record of solar activity. The most fundamental work an amateur solar observer can carry out, other than naked-eye work, involves recording how active the Sun is on any given day. There are a number of ways of measuring solar activity, and I intend to discuss the two main methods used by amateurs. The first is that used by the British Astronomical Association's Solar Section, which involves counting the number of 'active areas'. An active area (AA) is a spot, or group of spots, which is at least 10° from its nearest neighbour, and the number of AA's is recorded on every day on which the Sun is visible. From the number of active areas recorded each day the Mean Daily Frequency (M.D.F.) is calculated. (M.D.F. equals the total number of active areas divided by the number of days on which an observation could be made.) The M.D.F. is calculated typically at monthly intervals, and then when sufficient data is available the M.D.F. for the whole year is determined. If solar activity recorded in this way is obtained over a whole solar cycle of 11 years, then the overall curve of activity can be plotted on a graph with pleasing results. Taking the month of May 1978, for example, the following results were obtained by the author:

Total number of A.A's in the northern hemisphere = 89
Total number of A.A's in the southern hemisphere = 50
TOTAL = 139
Number of days on which counts were made = 24
Therefore, M.D.F. = 139/24 = 5.79
M.D.F. (North) = 3.71
M.D.F. (South) = 2.08

May 1978 happened to be an active month, but results obtained at solar minima are just as valuable.

The second method of measuring solar activity is that which is generally accepted as being the standard among astronomers throughout the world. This system was first introduced by the famous solar astronomer R. Wolf, and uses the following simple formula to give the so called 'relative sunspot number':

$$R = k \, (10g+f)$$

In this formula, R is the relative sunspot number, k is some constant, g is the number of groups, and f is the number of single spots. The problem with this method is the constant factor k which is extremely difficult to determine. This factor equals unity in the case of Wolf's own observations at Zürich, made with a telescope of 10cm aperture, and magnification of ×64. Therefore, if the telescope used is larger than 10cm aperture the value of k will be less than one, whereas smaller apertures will generally produce values of k greater than one.

For any records of solar activity to be really useful, details of observing conditions, the time of observation, and any other relevant information must be included. Observing conditions are not usually made using the Antoniadi scale with the Sun, instead remarks such as 'boiling violently' and 'good transparency' are more suitable. The effect of seeing the surface of the Sun apparently boiling is not, of course, a true solar phenomenon; it is caused by disturbances in the Earth's atmosphere. A good test of seeing conditions is to look for the mottled effect on the Sun's surface caused by the so called 'solar granulation'. A modest telescope under good conditions will show the granulation, which is caused by columns of very hot rising gas. The time of observations must be noted using the Universal Time scale (U.T.), which is the same as G.M.T.

When conditions in the Earth's atmosphere permit, it is often interesting to make drawings of sunspot groups and faculæ. For this purpose it is useful to attach to the screen in the projection-box a grid with a series of squares on it. This grid is used to position sunspots and faculæ with precision, and is made about 6 inches (15.2cm) in diameter. The grid must be aligned correctly, and this is achieved by allowing the image of the Sun's disk to drift across the field of view, so that a convenient sunspot or upper limb drifts precisely parallel to one of the horizontal lines drawn on the grid. The positions of sunspots and faculæ are then transferred on to another 6-inch (15.2cm) disk, which will be the actual drawing, using a suitable pencil for sunspots, and a yellow pencil for faculæ. The drawing should be transparent, so that an identical grid to that within the projection-box can be placed behind it for the purpose of drawing in the spot's positions.

Drawings of this kind are of no value whatever unless remarks about observing conditions, and the time of observation are included. If the observer then wants to determine from the drawing the longitude and latitude of sunspot groups, further complications are introduced.

The Sun's axis is inclined at an angle to the plane of the ecliptic, so that as the Earth moves in its orbit, the view of the Sun we have from Earth changes through an annual cycle. These variations are very complicated, and can be summarized using the following table:

P = position of N end of Sun's axis measured eastward from N point of disk.

B_0 = latitude of central point of disk.

Date	P	B_0	Date	P	B_0
	degrees			*degrees*	
Jan 6	0.0	−3.6	July 7	0.0	+3.5
Feb 5	−13.7	−6.3	Aug 8	+13.0	+6.2
Mar 6	−22.7	−7.25	Sept 8	+22.7	+7.25
Apr 7	−26.35	−6.2	Oct 10	+26.35	+6.2
May 7	−23.1	−3.5	Nov 9	+23.0	+3.5
Jun 6	−13.7	0.0	Dec 8	+13.5	0.0

Tables which give these figures at five-day intervals can be found in the B.A.A. Handbook, and using these in conjunction with specially prepared disks, such as the 'Stonyhurst Sun Disks' or 'Porter's Solar Disk', the positions of sunspots in solar longitude and latitude can be determined.

Drawings of sunspots and faculæ provide a valuable record of solar activity, and with good sequences of drawings, the individual life histories of spots from small pores to complex groups are obtained. Careful observation will also reveal such interesting phenomena as the 'Wilson effect', in which the penumbra of a spot is seen to be narrower on the edge of the spot nearer the centre of the Sun's disk. This is observed in some spots when they are close to the limb regions, and indicates that these spots may be depressions in the solar surface. Not all sunspots show this effect, and it can only really be attributed to some perfectly regular symmetrical spots.

With a great deal of patience an amateur solar observer will find rewards for his or her efforts. In a typical year the Sun will be observable on well over 200 days of the year from southern England. Keen observers find that they often have to sacrifice a breakfast to catch that vital glimpse of the Sun! The last solar maxima occurred in October 1968, so another maximum is due. By all accounts it should be better than the last maximum, which was about average; so now is the time to be observing the Sun!

Some Aspects of Martian Dust Storms observed during the Viking Mission

GARRY E. HUNT

1. *Introduction*

Although, by terrestrial standards, the Martian atmosphere is very thin, with a surface pressure of approximately 7mb, observations have shown it to possess a dramatically active meteorology. Perhaps the phenomenon, unique to Mars, that greatly modifies the weather and climate of the planetary atmosphere, is the dust storms which can sometimes be global in extent and last for several months.

Flaugergues suggested that ochre-coloured veils were the cause of certain obscurations of Mars during the period of 1796 and 1809, and his observations are some of the earliest records of the yellow clouds or storms. These yellow clouds, which appear with higher contrast when viewed in red light rather than in blue light, have been observed to spread into global storms which envelop the entire planet during most, if not all, apparitions when Mars is near the perihelion of its orbit. This occurs in the late spring/early summer of the southern hemisphere, when insolation is at maximum, and at a value of 20% more than average due to the orbital eccentricity.

Despite observational bias introduced by the restriction of data collection to periods when the angular diameter of the planet is large, and by the small number of seasons for which a multi-coloured photographic record exists, the data strongly suggests that global dust storms are an annual event on Mars.

Major advances in our understanding of these phenomena has come from the Mariner 9 and Viking space missions to Mars. It was perhaps fortunate for atmospheric scientists that Mariner 9 arrived at the planet in November 1971 when the atmosphere was engulfed by a major storm which was first observed by Earth-based telescopes in the previous September. At the height of the storm, dust was found to be uniformly mixed in the atmosphere to a height of nearly 50km, so that for a time, even the huge volcanoes of the Tharsis Ridge were hidden from view. This major atmospheric disturbance lasted for approximately four months.

Scientific instruments carried on the Viking orbiters and landers have monitored the atmosphere for an entire Martian year. During this time, large quantities of data relevant to the dust storms have been obtained. The two Viking landers, both situated in the northern hemisphere, are continuing to record changes in the atmospheric opacity, as well as the local meteorological data of winds, temperature, and pressures. The orbiters returned television pictures of atmospheric phenomena, observed brightness temperatures in 5 infrared channels, and mapped the concentration of water vapour in the atmosphere as a function of position and time.

Through this wealth of data, we now have a more complete picture of dust related phenomena which can be divided into three main categories: local dust storms, global dust storms, and background atmospheric aerosols as we discuss in this paper.

2. *Local Dust Storms*

During the orbiter mission so far, more than thirty local storms have been observed in the vidicon images and detected from the thermal infrared brightness temperatures. Dust storms can be positively distinguished from condensate clouds on vidicon images only when pictures are taken through both red and violet filters. These clouds may be detected in the infra red scans by their effect upon the brightness temperatures in the 7μm and 9μm channels. The dust is highly absorbing at the $9\ \mu$m channel, but does not exhibit a noticeable effect at the other wavelength. Therefore, the difference of T_7-T_9 indicates the presence of dust through altitude differences between the

Figure 1. Local dust storms are located by their latitude (ordinate) and aerocentric longitude of the sun Ls (abscissa). The time periods encompassed by the two global storms are also shown. The plot does not include all local storms seen within the geographical area and time period (from A. Peterfreund and H.H. Kieffer, J. Geophysical Res. in press).

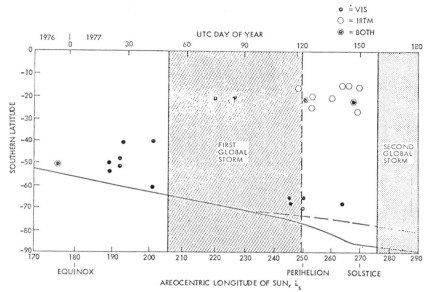

emission regions for the two wavelengths. For the most active local storms, this difference may be as large as 25K.

The storms observed by Viking were not distributed uniformly in time or on the surface of the planet. Even when biases introduced by constraints imposed by the area surveyed by the orbits of the spacecraft are considered, the local dust clouds are preferentially seen at restricted locations in the southern hemisphere during spring, Ls = 180° to Ls = 270°. This geographical distribution is shown in Figure 1 for southern hemisphere storms observed during this period. The figure clearly shows that the storms may be grouped into two distinct classes; namely, storms along the edge of the retreating south polar cap and storms in the 15-25° latitude bands. The second group was also concentrated in longitude, since all of the observed storms occurred in the Solis Planum region to the

Figure 2. A small dust storm observed on Viking revolution 248B at the edge of the south polar cap, which is just visible at the lower left. The time is Ls = 250° when Mars was at the perihelion of its orbit. These storms typically occupy an area of 10_5 km$_2$.

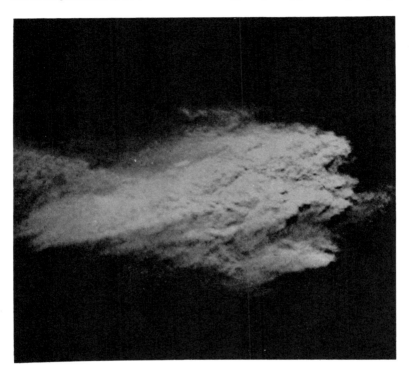

southeast of Arsia Mons and to the east of the Claritas Fossæ ridge. This distinction in the geographical location of the storms relates directly to the meteorological characteristics of the Martian environment that is responsible for their creation.

Most of the storms seen in very early spring occurred near the perimeter of the polar cap, and storms in this region continued into late spring. An example of one of these storms is shown in Figure 2. Four distinct storms were seen along the edge of the cap at this particular time, suggesting that a larger weather system was responsible for the small storms. The lobate edge suggests motion away from the edge of the cap

Figure 3 This storm is in the Solis Planum area at Ls = 228°. It is 1130LT. Dust clouds of this sort are extensively discussed by Peterfreund and Kieffer (loc. cit.). The western end of the Valles Marineris canyon complex appears to the north of the storm.

which is consistent with winds from the cap expected to be produced by CO_2 sublimation and by the large thermal gradient at the edge of the frost cover. Shadows indicate that most of the dust is in the lowest 10 kilometres of the atmosphere, and the structure in the clouds suggests convective activity. The winds from the cap may not be sufficient to lift the dust by themselves, and may do so when augmented by the general circulation winds at the edge of the cap.

Most of the Solis Planum storms occurred during the time period between the two global storms corresponding to mid and late spring. Figure 3 depicts a compact dust cloud seen in the region with a down slope motion to the southeast suggested by the lobate edge. Height indications confine these storms to the lowest 10 km of the atmosphere. The infrared measurements at 15μm which are sensitive to the atmospheric conditions at an altitude of 20 km, show no effects from the small storms and thus confirm shadow estimates. The region where these storms occurred exhibits unusually large temperature contrasts diurnally because of large thermal inertia contrasts in the area. The large thermal gradients produced would be expected to give rise to downslope winds which could produce the dust activity. The infrared diurnal observations indicate that the storms are most active during the early afternoon, and some data suggest that the storms regenerate during the night in the same location, thus producing a repetitive pattern of storms in the area. The location of the active dust storm area at 15° to 25° S is significant, since diurnal tidal effects are most sensitive to heating in this band of latitudes. The winds created in this way would be dominant mechanism for raising the dust in this region.

Other local dust clouds were seen in the northern hemisphere along the edge of the polar hood, although the fact that both space-craft had periapsis in the north limited observation opportunities there. Such storms were also recorded by Mariner 9, and they may be associated with the passage of cold fronts out of the north polar region. A very interesting storm was seen in the Chryse Basin quite close to the VL-1 site. It was observed on a group of consecutive frames which constituted a Phobos shadow observation sequence. These pictures led to determination of the speeds at which cloud features were propagating. The high 40-60 m/sec speeds deduced correlated with unusually high ground speeds observed by VL-1 at that time, but the storm had little effect on other lander meteorology instruments. This was possibly due to the arrival of the storm in late afternoon, when storm activity tends to be reduced.

3. *Global Storms and Background Aerosols*

Two global dust storms occurred during the Martian year

studied by Viking. Although the observational record is somewhat patchy, it appears that the occurrence of two global storms in one season is unusual. Furthermore, the observation of a global storm so very early in spring is very unusual, if not unique. The global storms affected most of the experiments performed by both the landers and the orbiters.

The first global storm was observed by the orbiter cameras at Ls = 204° in the Thaumasia-Solis Planum region. The storm resembled a large local storm at this early sighting, and was cearly restricted to a small area. Subsequent observations at two-day intervals showed a rapid spread of the storm throughout the southern hemisphere. This had some similarities with the origin of the 1973 storm, observed from the Earth. The orbiter pictures, which span several hours of local time, show strong convective activity which seems to peak in the early afternoon. Cloud shadows lead to altitudes as high as 30km for dust clouds during the early phases of this global storm. The storm spread very quickly to global proportions, with mean meridional speeds of 10 m/sec implied by comparisons of various data sets. The activity in the storm decreased as the storm spread and optical depth increased; most of the planet was obscured by a featureless haze for two to three months. When first seen at Ls = 275°, the second storm had already developed into a large disturbance (see Figure 4). It also seemed to have originated in the Thaumasia region.

The clear Martian atmosphere observed at the Viking lander sites during the primary mission, which is coincident with southern winter, has an optical depth at visible wavelengths of between 0.3 and 0.5. Although condensate fogs may have contributed to the larger values obtained in the early morning lows, the background contribution due to aerosols of about 0.3 is indicated. This opacity is sufficient to create the pink sky observed in the lander images, since there are insufficient air molecules to have any scattering effect on the sunlight. A gradual build-up in opacity to about 0.9 preceded the arrival of the first global storm, which then caused the optical depth to rise rapidly to about 3.0. Following a gradual decay back to an optical depth of approximately 1.0, the second dust storm

Figure 4. A mosaic of the Martian southern hemisphere completely shrouded in dust during the second global storm Ls = 356°.

caused another rapid increase to ~ 3.0. These rapid changes were observed at both landing sites within three days of the first observations of the storm, although the most significant effects were noticeable at the Chryse site.

The global dust storms had a large effect on the climate and weather patterns of the entire planet. The infrared observations made in the 15μm band during the second storm showed that the atmosphere over the north pole was extensively warmed by the intrusion of dust being mixed in the atmosphere at high elevations. The overall effect was to increase the temperature by 50-60° from ~ 130K in about three weeks.

Since the north pole receives no direct solar radiation during this season, this rapid response of the atmosphere is due to the atmospheric motions. In addition, the downward radiance from the warm dusty atmosphere would be expected to inhibit CO_2 condensation and this was probably responsible for a pulse in the total pressure observed at the landers.

The meteorological experiments at the lander sites registered effects of dust on the diurnal variations in the temperature and pressure. The diurnal pressure ranges exceeded 15% of the mean pressure at the VL-1 site immediately after the onsets of the dust storms. In addition, the pressure range variations showed a similar structure to those of the opacity changes, although the decay of pressure after the main dust storm events was more clearly defined than was the corresponding behaviour of the opacity during these same two periods. After the onset of the first dust storm, the temperature range at both sites was much lower than during the early mission, as a consequence of the increased opacity and decreased net radiation at the surface. The day-to-day variability of the temperature range was much smaller during the peak of the main dust storm, indicating the uniformity of the observation at that time. While the arrival of the storms at VL-1 were also accompanied by significant increases in the local wind speed, no such changes were observed at VL-2. During the daylight period of VL-1 sol 209 (Ls = 205°), the wind speed rose steadily, attaining an hourly average value of 17.7 ms^{-1} at about 1900LLT with numerous gusts peaking at 25.6 ms^{-1}; values significantly larger than any previously recorded at either site. The speed of the higher gusts at the 1.6m instrumental height corresponds to a friction velocity of about 2.0 ms^{-1} for an assumed aerodynamic roughness length of 1cm. It thus appears possible that surface material could be raised at the landing site, and such local effects would contribute to the rapid onset of the first storm at VL-1. The wind directions during the period followed the same diurnal pattern as on previous sols, switching from westerlies to easterlies in the early afternoon, implying an intensification of the tidal pattern rather than a response to a synoptic pressure system. The arrival of the second storm was marked by more gradual changes, with average wind speeds of 13.9 ms^{-1} and gusts of 21.2 ms^{-1}.

At the second site, VL-2, the arrival of the first storm was less pronounced as a consequence of the larger horizontal temperature gradients at this latitude, and appreciable thermal advection associated with synoptic systems moving across the site. Since the surface winds remained predominantly northerly, it suggests that the dust moves in at relatively high altitude. The maximum wind speed of 13.6 ms^{-1} does not seem sufficient to raise the surface material. At the time of the second storm, the minimum surface temperatures in the VL-2 region reached CO_2 condensation level, so that we would not expect local raising of the dust from a frost covered surface.

4. *Dust Storms and Martian Climatic Change*

These Viking observations clearly show the close connection between the global meteorology of Mars and the dust storms. In particular, the link between oscillations and the storms is particularly significant. We have already seen that the temperature gradients, adjacent to the edge of the retreating south polar cap, set up a baroclinical zone, which produces winds of sufficient strength to create local storms well in advance of perihelion. Tidal effects, which will be strongest at perihelion, will be dominant in creating the storms observed in the Solis Planum region. However, we still have to account for the spreading of the dust. The close connection between the tidal amplitude and opacity supports the idea that global tides, excited by increased solar absorption prior to the storm, are the major dynamical factors in storm genesis. The tidal pressure variations drive wind systems of planetary scale and when they reach sufficiently large amplitude, they are capable of raising dust over very wide areas. Dust storms are therefore a unique Martian meteorological phenomenon.

But what factors govern the lifetime of these major storms? The gradual decrease in the opacity, reducing the solar heating of the atmosphere, and therefore the atmospheric tides, is due to a gradual fallout of dust. Why should this sudden storm shut-off occur just at the onset? Several factors could be involved, but perhaps the most important of these is the cut-off of the slope induced by the thermal wind contribution to the total wind. As the atmosphere becomes uniformly dusty, the differential heating associated with the slopes would be eliminated, and with it the main dynamical forcing of the storm.

In addition to the direct effect upon the meteorology, dust storms have a major impact upon the climate of Mars. They directly impact and interact with the planet's water budget. The behaviour anticipated for the water vapour in the southern hemisphere was inhibited by the occurrence of both dust storms, with the result that the moist conditions, which occurred during the northern summer, did not recur in the southern hemisphere. At the polar region, some dust may indeed become mixed with ice deposits, resulting in a 'dirty polar cap'. The dust particles act as cloud nuclei for the condensing atmospheric CO_2, which, as the particles grow, precipitate out of the atmosphere forming a surface layer of mixed composition. This mechanism also accounts for the clarity of the atmosphere in polar regions after the storm season.

A further major aspect of the global dust storms is the rôle they play in modulating the seasonal storage of volatiles in the polar caps. Seasonal changes of the polar caps are sensitive to the insolation changes in the atmospheric heating, caused by the dust, and will therefore be reflected in the condensation of the polar deposits. Indeed, the dust storm activity, observed during the Viking period, has had a direct effect on the south pole, whose retreat was considerably slower than usual. The magnitude of the retardation at summer solstice is estimated to be 12.5° in Ls. Furthermore, it would appear that CO_2 remained on the surface at the south polar cap throughout the summer season observed by Viking. There is little doubt, therefore, that the periodic variations in the frequency of the global dust storms are reflected in the laminated structure of the polar deposits.

As time progresses, we may expect the Martian dust storms to denude the middle southern latitudes of dust, exposing the bedrock and leaving behind deposits of darker, less weathered materials, which are probably the classic dark areas observed from Earth. These are shaped by the winds moving across regions of differing surface properties, and enhanced by the local topography. Over a period of 50,000 years, the perihelion subsolar latitude of Mars will vary between ±25°. Consequently, the dark areas of Mars may then also migrate around the planet within the same period. At past epochs, the perihelion dust storms may have originated in the Chryse basin, which would then have been stripped of the brighter deposits.

The Martian dust storms, created by a vigorous interaction between the planetary surface and the atmosphere, are therefore a special characteristic of the planet's meteorology. Their effect upon the planetary environment, particularly at polar regions, is so significant that the laminated terrain may directly indicate the past storm frequency. However, our understanding of these phenomena has been significantly advanced through the measurements made during the highly successful Viking mission, which has monitored the Martian atmosphere through a full year.

Acknowledgements

It is a pleasure to thank my colleagues, associated with the Viking project, for many helpful discussions. In particular, Professor H.H. Kieffer, Professor C. Leovy, and Professor P. James. My participation in the Viking project and this research is supported by the Science Research Council (UK).

A major discussion of Viking studies is contained in the following references.
1. *J. Geophysical Research* **82,** September 1977.
2. *J. Geophysical Research* May, 1979.

Some Properties of Open Clusters

ÅKE WALLENQUIST

Stars have a common tendency to appear in groups. Binary systems, i.e. systems consisting of two stars, are very frequent, so that every second or third star may be a binary. Triple and multiple stars are also rather common phenomena.

If we look carefully at the sky, we can see a few larger groups of stars even with the naked eye – such as the Pleiades in the constellation of Taurus, and Præsepe in Cancer. These groups or clusters are relatively near, so that individual stars may be distinquished with the naked eye. In more distant clusters, no individual stars can be seen, so that the clusters appear as diffuse nebulæ resolvable into stars only with optical aid.

A stellar cluster consists of a number of stars whose members are physically associated, forming a small, isolated stellar system. Clusters are of two different types: these are known as *globular* and *open* (or galactic) respectively. The former are enormous conglomerations of closely packed stars, which even in large telescopes look like diffuse spheres. They may contain hundreds of thousands, or perhaps millions of old stars. The globular clusters lie at very great distances from us, and surround the Galaxy as a swarm of satellites.

The open clusters, on the other hand, are typical Milky Way objects, and are found in the plane of the Galaxy – in the disk structure as well as in the spiral arms. Therefore, they are also known as galactic clusters. They are much smaller and looser systems than the globular clusters, and contain far fewer stars. Moreover, the members of the open clusters are mainly young stars of the same age as the stars in our own region of space. In this article, some properties of the open clusters will be discussed.

At the present moment more than 1200 open clusters are known, but the total number within our Milky Way system is probably more than 30,000. They differ with regard to the numbers of stars, the concentration of the stars toward the centre of the cluster, and in the range of magnitudes and colours of the member stars, as shown in the Hertzsprung-Russell or H-R diagrams of the clusters. The H-R diagrams of open clusters also differ from those of non-cluster stars, since the cluster stars occupy only a part of the Main Sequence in the diagram. Through the studies of the H-R diagrams, it is possible to estimate the ages of the clusters. The youngest are about a million years old, whereas the oldest have ages of about 10,000 million years.

In general, an open cluster consists of some hundreds or thousands of stars contained within a spherical volume from 5 to 10 light-years in radius. Most of the known open clusters lie between 1000 and 5000 light-years from us. The most remote open cluster so far measured is about 20,000 light-years away.

Since the cluster dimensions are very small in comparison with their distances, the distribution of the apparent magnitudes of the stars is directly related to the absolute magnitudes, allowance being made for distance. It is therefore possible to determine the distance of a cluster by means of its H-R diagram as soon as the absorption of light in the direction toward the cluster is known.

Since all the stars of a cluster are formed within the same cloud of dust and gas, all members of the cluster are of the same age, as well as chemical composition, so that the differences between the stars are due mainly to their differing masses. The study of open clusters is therefore of the greatest importance in increasing our knowledge of the birth and evolution of the stars.

Investigations of the structural properties of open clusters are based mainly on studies of the distribution of the stars in space. A photograph of a cluster is a projection, against the sky, of the cluster stars as well as all other stars which happen to lie in the direction of the cluster. In order to eliminate the background and foreground stars which do not belong to the cluster, i.e. the 'field stars', we must study the variations of surface density – that is to say, the numbers of stars per unit of suface area – at various distances from the centre of the cluster. Then, using statistical methods, we can identify the field stars. This leads on to a calculation of the number of stars lying within a cubic parsec (there are several ways of doing this, but to deal with them here would be too much of a digression). Thus, we can derive space density curves which show the real distribution of the cluster stars.

By means of the space density curves, it is also possible to determine the relative mean masses for various types of stars, assuming that an open cluster may be regarded as a gaseous mass in isothermal equilibrium, in which case the Maxwellian law of distribution is valid. In the Maxwellian distribution, the most massive stars show a stronger concentration toward the centre of the cluster than for stars of lower mass. In Uppsala, this method of determining stellar masses has been applied to several open clusters, and the results are in good agreement with the masses as determined by studies of binary systems.

An open cluster is subjected to the influence of various forces of dynamical origin. Literally speaking, the dynamical state within a cluster will be governed by the laws of chance and of gravitation. Because of the total gravitational pull of the cluster, the individual stars will describe orbits around the centre of the cluster. Sometimes, two stars will move so close together that their orbits will be disturbed. After such an encounter, the star with the smaller mass will be left with a larger velocity than that of the more massive star. Therefore, it will move to a greater distance from the centre of the cluster. Such stellar encounters are probably very rare, and for a given star may be expected only once over a period of several million years. Yet a star's lifetime extends over thousands of millions of years, and, therefore, such random encounters will little by little lead to a state of equipartition of energy of the same kind as that prevailing in a gaseous mass in isothermal equilibrium.

However, it may also happen that some cluster stars will be

Figure 1. Messier 8 in Sagittarius, an open cluster involved in gaseous masses. (Palomar and Mount Wilson observatories)

given velocities so large that they exceed the total gravitation of the cluster. Due to such encounters, stars may escape from the cluster. Such an escaping star will take away a little of the total energy of the cluster, so that the cluster will tend to contract. Over a sufficiently long period, the numbers of stars in a cluster will steadily decrease, and the cluster diameter will become smaller with increasing age. The final stage of an open cluster may in fact be in the form of a multiple star.

With loose clusters, the tidal force as well as the rotation of the Galaxy will disturb the cluster stability, so that the clusters will finally be dispersed and their members mixed in with the general field stars.

As already mentioned, it is very probable that a cluster is born inside an interstellar cloud of gas and dust. Young clusters may still be involved in clouds of luminous gases. (Figure 1). With older clusters, no gaseous matter remains, but interstellar dust shows up as reflection nebulæ in the neighbourhood of bright stars (Figure 2). However, in most clusters it is impossible to detect directly the existence of dark matter whether inside or adjacent to the cluster (Figure 3).

In order to investigate the existence of dark matter, two different methods are at our disposal. The first is based on the study of colour excesses in the cluster region; the second depends on star counts.

During recent years, attempts have been made at the Astronomical Observatory in Uppsala to determine dark matter in and around about 150 open clusters, using the star-count technique. The counts were made from Palomar Sky Survey prints, and, for southern clusters, on plates secured at the European Southern Station in Chile. The material has been only partly analysed as yet, but some preliminary results have been obtained.

On each cluster, the counts were made by using a transparent réseau with large numbers of equal-sized squares. The réseau was placed over the cluster print and counts were made by using a binocular magnifier with a magnification of ×20. Within each square, all stars clearly visible were counted. The numbers were then transferred to maps which showed the

Figure 2. The Pleiades, an open cluster where the dark matter appears as reflexion nebulæ. (Uppsala Observatory Kvistaberg Station, the author's photograph)

surface distribution of the stars in the regions concerned.

In order to eliminate (as far as possible) the influence of the systematic increase in the numbers of stars toward the centre of a cluster, the cluster region was divided into a number of

Figure 3. Messier 11 in Scutum, an open cluster containing dark matter. (Lick Observatory, the author's photograph)

concentric rings. Within each circular region, the average number of stars per réseau square was computed, as well as the deviation of the numbers in the individual squares from this average.

The fundamental assumption underlying this investigation is that the squares with deficiency of stars are affected by interstellar absorption due to dark matter. In order to avoid the influence of statistical fluctuations in the projected stellar distribution within the cluster regions, we used only those squares in which the number of stars was at least 25 per cent less than the average; these were known as 'deficiency areas'. The variation of the surface densities of the deficiency areas according to distance from the centre of the cluster was then studied graphically.

Within many of the clusters, dark matter could be traced. There was in fact evidence for the existence of dark matter in more than 60 per cent of the clusters investigated. Within about 3 per cent., absorption zones were found, and also indications that the distances of these zones from the cluster centre increased with the age of the cluster. It seems as though the dust remaining after the forming of the stars has been driven away from the cluster, and is now, in many cases, far outside the cluster itself.

The T Tauri Stars

MARTIN COHEN

We have a time machine and we are going to travel backwards to the early history of our Solar System, to a time before there were planets as we know them, to a time when the Sun was not at all as it is today. The question we are trying to answer is simple: what was the Sun like when it was only a million years old, or ten thousand? (It is now some 6000 million years old.)

Things in our universe change very slowly by comparison with the life of an astronomer, and we cannot travel through time yet. Consequently, what we do in order to study the early evolution of the stars is to look around our galaxy for a place where stars are still being born. We find different stars at different stages of their evolution, and by placing them in a time sequence we can build up an impression of the way in which a single star would alter if we could watch it for æons. Let us restrict our discussion to stars of mass comparable with that of our Sun. The nearest birthplace of stars at present is the complex of dense, dark dust clouds in the constellations of Taurus and Auriga, some 500 light years away. In Taurus lies the variable star T Tauri (which was not discovered by one Thomas Tauri, despite the rumour!), a tenth-magnitude star which occasionally exhibits variations of about a magnitude or less, and lies within the dark clouds. What are the characteristics of the class of T Tauri variable stars, stars which resemble T Tauri itself in their general nature, and why do astronomers believe that these stars are young?

Firstly and obviously, the only places in which T Tauri stars are found are in dense dusty clouds, just the sort of locations to seek giant collapsing clouds of gas and dust that will one day form stars. Secondly they are gregarious creatures, rarely being found in isolation but rather in crowds of tens or more stars in each region of a dark cloud complex. Thirdly, they reveal stellar gaseous atmospheres indicative of temperatures of a few thousand degrees, like those of normal stars with mass from much less than the sun to two or three times that mass. But their optical spectra also reveal a pattern of many bright gaseous emission lines that tells us that in the atmospheres of these stars there are hot glowing streams of gases, of hydrogen, oxygen, nitrogen, sulphur, iron, to list only the most abundant elements.

These lines are very strong, in fact uncommonly so for normal, mature stars do not show them at all. In the blue region (and ultraviolet too) their spectra deviate from normal spectra in showing far too much light, as if there were some extra in their atmospheres not found in normal stars. Some members of the class exhibit rapid variations in total light output, as fast sometimes as one magnitude brightening or dimming within a day. All these characteristics are grossly abnormal compared with most stars that we know.

There is more to the story, however. In the past decade it has been possible to make observations in the infrared region (very long wavelength radiation, some four to forty times the wavelength of ordinary light). The importance of infrared radiation is that it often signifies heat from dust grains at temperatures of hundreds of degrees (compared with the thousands of a typical stellar atmosphere), or perhaps radiation from other less common processes. The T Tauri stars are rich in infrared radiation too; in fact, the only spectral region in which they have thus far been found to be even approximately normal in their continuous (not line) radiation is the narrow visual window!

Can we synthesize all these facts into a consistent picture? A huge cloud of gas and dust grains begins to contract, becoming tighter and tighter in its centre. After hundreds of thousands of years a core of material is formed that is hot enough to trigger nuclear reactions and make the gas cloud self-luminous. A star is born. Subsequently this infant star evolves by accreting the remaining infalling cloud of gas and dust that surrounds it. This cloud is responsible for absorbing much of the starlight in the early phases of evolution, and this energy reappears as infra-red emission from the dust heated by starlight. Consequently it is vital to observe the young T Tauri stars in the infrared as well as in the optical region to know accurately how bright they really are.

This much is known reasonably securely about these stars. However, there are many mysteries along the road of early stellar evolution. Many very faint T Tauri stars that have been studied in detail reveal quite well-defined stellar atmospheres at extremely early ages, of order a few tens of thousands of years! Yet the dust cloud would be expected to veil the young stars for many hundreds of thousands of years. So it has been necessary to postulate that these stars turn on a wind, a highly energetic stream of gas moving at appreciable speed, that scours clean the immediate vicinity of the star. This wind renders the stars visible at a much earlier phase than would have been expected. It is not a ridiculous thing to postulate, however, for it is characteristic of older T Tauri stars to show evidence for such a wind, although not one as energetic as would be needed in the earliest phases of evolution. Integrating this into the picture one would say that, as a T Tauri star ages, all the associated abnormal phenomena become less pro-nounced. As the star ages, the wind becomes less active, the amount of luminosity converted into infrared emission dimin-ishes, the star becomes more and more like a normal star of the appropriate mass.

Another way in which it is fruitful to view T Tauri stars as extreme versions of normal stars, like the Sun, is to argue that the rapid variations that are seen in total light (and also in the optical emission line spectra) are due to flare activity. Now, everyone knows that the Sun undergoes flares, gigantic uphea-vals of its surface in local areas, often related to sunspot regions, in which large masses of gas are thrown from the surface to immense heights. This gas glows and is responsible for the famous prominence ejections when seen in the correct part of the solar surface. In like manner it may well be that T

Tauri stars undergo flare activity too, but on a much more energetic scale. There is already evidence that some T Tauri stars throw material to great distances from which some returns to the stellar surface. This constitutes a type of stellar recycling of material. It equally might by that the T Tauri winds are not so much an organized global activity as a series of superflares. As the star ages, this activity too would decline until a relatively sedate star like our Sun was produced, still flaring but producing only a tiny fraction of its total surface energy through this mechanism.

Actually there is another mystery to inject at this time in the story. If young stars are surrounded by extensive clouds of material which we believe must fall into the protostars, why do we never see evidence for such matter in the spectra? Why is all the gas that we do see either leaving the star, or falling back after having left previously? This argument suggests that if winds are turned on then this must occur at extremely early phases, so that as soon as the star is optically visible to us, the shell of gas has been dispersed.

Slowly over the past decade or so, we have been trying to push back the age of the youngest star that we know. Some years ago that age might have been about a million years, then three years ago the age receded to perhaps three hundred thousand years (represented by the strange T Tauri star HL Tauri; see *Yearbook of Astronomy 1977*, 'Ice in Space'). Now we believe that we have identified a T Tauri star in Cepheus that is as young as only 30,000 years! The properties of this star may hold significant clues as to the nature of even younger stars and the attempt to regress even further will continue, from a synthesis of optical, infrared and radio observations.

We noted that T Tauri stars are gregarious. Now, suppose that you are a giant association of gas and dust and stars in Taurus, or Orion, or Cepheus. You are given, let us say, ten thousand solar masses of gas and dust from which to produce stars. How will you choose to do it? Will you make 1000 stars, each ten times the mass of the Sun? or perhaps 10,000 stars like the Sun? or will you manufacture large numbers of tiny stars? Indeed is there any reason why each association should make stars according to the same recipe for mass distribution? Are there even different mechanisms for making stars different (high and low) masses? These are the great riddles towards whose solutions both observers and theoreticians are working. For example, the Taurus clouds abound in T Tauri stars, stars which seem to be mass comparable with that of our Sun. But in Orion there are many O and B stars, stars of high temperature and mass much greater than our Sun's and correspondingly fewer T Tauri stars. Why? Another clue to this problem may come from recent radio observations of clouds of molecules like carbon monoxide. Radio astronomers too have looked at the dense, dark places in the sky where star formation could be occurring. They see much general gas, but this background is stippled with a very small number of bright, very dense, spots. In these spots the enhancement of gas density is many, many times that in the background. Sometimes there are associated infrared sources, and, less frequently, fuzzy faint stars that reveal the characteristics of young stars to the spectroscope. It appears that a giant complex of clouds may decide to produce stars at different places at very different epochs. The fragmentation process may thus affect only a small part of a big dark cloud at a given time. Again there are many unanswered questions. For example, is the lack of high mass stars in Taurus due to the fact that the lighter, T Tauri, stars form first; or are there really more massive stars but deeper inside the clouds where light does not escape, and even infrared is impeded?

It should be apparent that in order to study star formation one must proceed on two very different planes; one must retain the details of the big picture, namely the associations or giant groups of stars, whilst not losing sight of the intimate details of how a young star produces its energy and its optical emission lines (these are vital clues to the temperature and density of the surrounding gas from which the star was born). Often in this field can one generate an apparently excellent experiment that produces results which merely reveal more complexities. As an example, a colleague and I some three years ago set out to monitor a sample of brightish T Tauri stars for six nights continuously, observing each star once or twice a night. We observed five optical wavelengths, spanning the ultraviolet to near-infrared range, then six long wavelength infrared colours, and finally we measured the stellar wind activity using an optical emission line. It was our hope that if any stars varied while we were watching them, the correlations between optical, infrared and wind properties would reveal vital hints as to the processes at work. About fifteen stars did vary. But we found every form of behaviour! Some stars dimmed optically and brightened in the infrared; some dimmed in both; some never changed in one characteristic but leaped all over in their other two. There was no uniform pattern at all! Faced with such a situation it is almost impossible to intuit the real physical behaviour of the stars, and to test different theoretical models. Of course that experiment did suggest another one. We have now been watching the optical spectra of about thirty 'pet' T Tauri stars for some two years, visiting each star at least once a night for periods of three or four nights each winter month. The spectra do vary, but only ten or so, out of the total of perhaps 100 emission lines, change in intensity. The clue to the winds may well hinge upon the correct interpretation of this set of data. In order to pin down the nature of our hypothesized flares I have also been looking at these same stars at radio wavelengths.

It is this mixing of techniques that distinguishes the astronomy of the 1970s from that of earlier times. It is an extremely exciting time to be around astronomy, especially when one is grappling with a vast problem such as star formation. Yet there is an almost classical, pioneering air to the field now, despite the new machines and techniques. For we have accumulated some data on each of about 500 young stars now, mostly of T Tauri type. The very process of observing is fascinating but the true excitement comes after, in analysis of the data, in searching for patterns of behaviour that will unify the 500 young stars into a small number of selected patterns. One generates a hypothesis, tests it according to the existing data, predicts new patterns, reobserves the stars in a different spectral region, and so on. One may be obliged to use a radio telescope, or an infrared telescope, of course, in order to test a theory based upon optical data. Slowly, then, we proceed in our interpretation of the enigmatic T Tauri stars. So far they are proving to be noble adversaries!

The Stripping Galaxy

DAVID A. ALLEN

This tale began some years back. Tim Hawarden was passing through England on his way to spend a couple of years at the U.K. Schmidt Telescope Unit in Australia. He dropped into my office in the Castle to discuss various astronomical matters.

'You'll be in an enviable position at UKSTU,' I commented, 'with access to most of the photographs of the southern sky. You'll have first view of many of the discoveries that telescope will make.'

Tim nodded: I wasn't the first to have said this to him. I pressed my point, however. On the same mountain as the Schmidt stands the Anglo-Australian Telescope, known affectionately as the AAT. This is probably the best telescope in the world at the moment, and certainly an ideal instrument to follow up discoveries made on the Schmidt. 'You should apply for AAT time to follow up those discoveries,' I added. Tim thought a moment, then wrote something on the back of his hand. The discussion turned to other matters.

Some months later, I too was installed in Australia, based at the AAT's headquarters in Sydney. A letter landed on my desk: within it was a Xerox copy of an application for AAT observing time. It bore Tim's name, together with Andy Longmore and Russell Cannon, both of the U.K. Schmidt Telescope Unit, and me. The team was formed.

Our application went in with more than fifty others. A committee of wise and learned elders of the astronomical community pondered and pontificated on these applications. For every night available on the telescope in the relevant three-month period, four had been requested. This was stiff competition indeed, and we counted ourselves lucky that the committee decided to award us a couple of nights on the great telescope.

We prepared our programme carefully. The Schmidt was undertaking a photographic survey of almost 20,000 square degrees of the southern sky, and a great wealth of unusual and interesting-looking objects were turning up on the photographs. One of the objects we selected was NGC 5291, in Centaurus.

What had excited us about this galaxy was a string of faint, fuzzy objects stretching to north and south from it. The photograph (Figure 1) shows these well, and is enlarged from the original Schmidt photograph. NGC 5291 is the oval galaxy in the middle of the picture. Just lower right of it is a small companion which has no catalogue designation, and which we dubbed the Seashell because of its whelk-like appearance. The fuzzy knots define a slightly curved line on this picture, convex to the left.

Our nights on the AAT were clear. We managed to secure spectra of NGC 5291 and the Seashell, and of some of the fuzzy knots. Spectroscopy is the most powerful technique available to optical astronomy. It taught us a great deal about the NGC 5291 system. The fuzzy knots, we learnt, are great clouds of gas illuminated by hot, young stars. Their ages are probably a few million years. They form a system of satellites orbiting NGC 5291.

The dimensions of the system amazed us most. Astronomers measure distances in light years. In one second, light travels 300,000 kilometres, most of the distance to the Moon. There are over thirty million seconds in a year, so a light year is about ten million kilometres, a distance quite unimaginable to those of us constrained to dwell on earth. Yet the distances involved in galaxies are measured in thousands of light years. Our own galaxy is some 70,000 light years across, but is dwarfed by the collection of satellites to NGC 5291, which spread more than 600,000 light years from end to end. They are orbiting NGC 5291 at a velocity of nearly 400 kilometres every second, yet so great are the distances they travel that each would take more than one thousand million years to complete one circuit. From these facts we could determine the mass of NGC 5291, and were surprised to find a figure of three million million times the mass of the Sun. NGC 5291 is, in fact, the second heaviest galaxy known, after M 87 in the constellation Virgo.

We also learnt that the Seashell is an interloper, quite irrelevant to the outlying knots. NGC 5291 lies at one edge of a cluster of several dozen galaxies. The Seashell is another member of the cluster which happens to be making a close pass of the larger galaxy, and which will in due course move away to pastures greener. Two other members of the cluster are seen on the photograph, lying to upper right and lower left. These are normal-sized galaxies.

The system of satellite clouds to NGC 5291 intrigued us. Why are they there? Had some great event in the history of the galaxy caused their formation? If so, what? We couldn't decide: we needed more information.

Andy Longmore made the next breakthrough. He was using the 64-metre radio telescope at Parkes to measure the amount of neutral hydrogen gas in nearby galaxies. On a hunch, Andy turned the great dish on NGC 5291. This was a long shot, for the distance to NGC 5291 is about 270 million light years, well beyond the range at which the neutral hydrogen normally present in galaxies can be detected. After 15 minutes of monitoring the galaxy, Andy thought he was detecting it. He pressed on. The next observation confirmed the detection. Andy made some calculations. Astronomers rarely, if ever, make their calculations on the backs of envelopes, as they are fabled to do, for they are invariably surrounded by mountains of chart recordings, computer printout and notepads. Andy's quick calculation showed that NGC 5291 contains one hundred thousand million solar masses of gaseous, neutral hydrogen. This is the largest single mass of neutral hydrogen ever found.

The hydrogen fills in the space between the central galaxy and the nebulous knots. Thus we learnt that NGC 5291 is not just the fairly normal-sized galaxy seen on the photograph, but a huge system ten times bigger than our own galaxy. Its outer

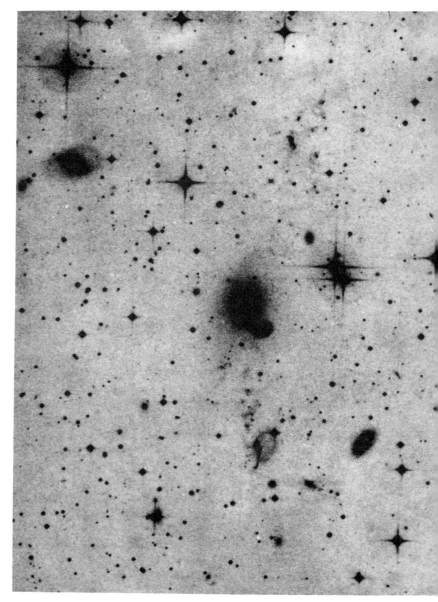

Figure 1 (opposite). Photograph, enlarged from the original Schmidt one, of NGC 5291 and the string of faint, fuzzy objects stretching from north (top) to south.

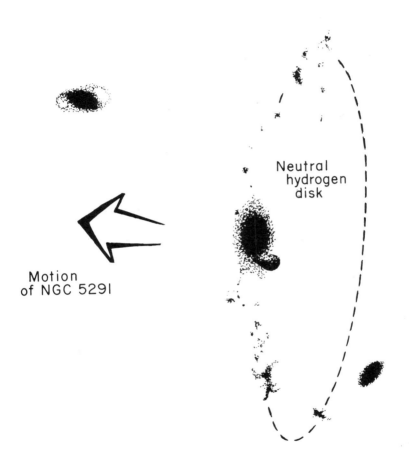

Neutral
hydrogen
disk

Motion
of NGC 5291

Figure 2 (above). Sketch showing situation of NGC 5291 undergoing gas stripping. Only galaxies are shown; foreground stars are omitted.

regions are not seen because few stars exist there. Some stars have, however, formed: they illuminate the gas to produce the nebulous knots. Had these stars not formed, our attention would not have been attracted to NGC 5291.

By mapping the distribution of hydrogen, a difficult business, we were able to confirm our suspicions: the gas lies in a flattened disk, just as do the stars in a spiral galaxy. The visible galaxy forms the hub of this disk. We view the disk from nearly edge-on, so it takes on a slender cigar shape. The nebulous knots lie around the rim of the disk, but occupy only one half of the rim, to the left in this photograph, This appears to be the rim nearest to us. Over the last few million years, something precipitated the formation of stars only around this rim. To understand what, we have had to delve into the literature on clusters of galaxies.

It has long been known that clusters of galaxies contain more material than just the galaxies we see. Between them is a tenuous ocean of hydrogen. Unlike the neutral hydrogen within galaxies, however, that between galaxies is very hot, ionized. It cannot be detected by radio telescopes, but only by X-ray detectors flying high on satellites above the Earth's atmosphere. The gas is heated by the galaxies themselves. Each member of a cluster of galaxies is constrained by the universal force of gravity to journey through the cluster at velocities of about one million kilometres per hour. When stars and gas meet, the effect is minimal, for the stars move through the gas as fish through water. But gas is sticky stuff. The gas within a galaxy collides with and sticks to the cluster gas, as if the two were porridge. Two phenomena result. First, the galaxy's gas is swept out and left behind. Second, the cluster gas becomes heated to a temperature of several million degrees. Its temperature is raised so high because the vast energy of the moving galaxy can only be released as heat. The hot gas radiates at X-ray wavelengths, and is recorded by satellites. The removal of the gas from the galaxies is a process known as gas stripping. Radio astronomy provides evidence of gas stripping: galaxies in clusters contain less neutral hydrogen than galaxies in the field.

Because NGC 5291 contains so much neutral hydrogen, it can never have been deep within the cluster with which it is associated. But we can be fairly certain that it is now involved with the cluster, and is thus approaching for the first time. Perhaps NGC 5291 was formed at the outer edge of the cluster and has taken several million years to fall in. It is now moving roughly towards the centre of the cluster, which lies in front and to the left of NGC 5291, off the edge of the photograph. This motion also means that the disk of neutral hydrogen is entering the cluster almost edge-on, like a giant frisbee. Over the last few million years NGC 5291 has encountered the cluster gas, and is therefore undergoing gas stripping. The gaseous disk is already falling behind, and the hub of stars which forms the visible galaxy has overtaken it almost to the rim. The situation is sketched in Figure 2, which shows only the galaxies and omits the foreground stars.

NGC 5291 is of singular importance because no other galaxy is known to be undergoing gas stripping, a process otherwise only predicted on theoretical grounds. Because we are witnessing the galaxy in the act of stripping, we see aspects of the process that theoreticians had not predicted. Most important of these is the formation of the nebulous knots. They are caused by the leading edge of the frisbee of neutral gas becoming squeezed in its collision with the ocean of cluster gas. Where gas is squeezed to higher density, stars form in it. These stars have congregated into a collection of knots. But again the scale of the system is remarkable, for the knots themselves are as large as many normal galaxies. Our own galaxy is accompanied by the Magellanic Clouds: these are of comparable size to the numerous knots of NGC 5291. Thus we have unexpectedly found a mechanism for the genesis of new, young galaxies in clusters.

When we first stared at the photograph of NGC 5291 and its associated system, we had no idea of what had occurred there. Recently, some European astronomers also noticed the system, and published a photograph with the caption 'Whatever happened to NGC 5291?'. It took us fifteen months to work out, to our satisfaction, what actually had happened there, and a further nine months passed while the paper we wrote journeyed through the laborious process of publication. Only in mid-1979 did our paper surface before the critical eyes of the world's astronomical community; as yet there has been no time for response. Other astronomers may place a different, a better interpretation on the facts. This Yearbook article is thus topical, but also biased by our own thinking. Whatever posterity considers really did happen in NGC 5291, there is no denying that it is a remarkable system. I count myself lucky to have been involved in the discovery and study of the galaxy.

Gamma Ray Astronomy

A. W. WOLFENDALE

Most of the facts we have about our Galaxy, and the Universe beyond, come from studies of the electromagnetic radiation that is falling on Earth. Until the 1940s the part of the spectrum accessible to observation was simply the optical window – the part visible to the naked eye – but technical improvements, many of them associated with techiques developed during the War, have since then allowed other parts of the spectrum to be examined.

The radio band was the first to be opened up and many important discoveries have been made in the field of radio-astronomy: principally the observation of radio galaxies, quasars, and pulsars. In each case the objects had not been predicted beforehand and each corresponds to a very energetic object on the astronomical scale. Continuing to shorter wavelengths, the remaining parts of the spectrum have also been exploited; the infrared, ultraviolet and X-ray region. Again, the sky looks very different in these regions, different objects usually coming into prominence as we proceed from one wavelength range to another. The ultimate wavelength band to be conquered is that of gamma rays (i.e. quanta with energy above 100 keV – one fifth of the rest mass of the electron – or so) and this article deals with the discovery of cosmic gamma rays and the view through this window. In so far as the author is a cosmic ray physicist, inevitably attention will be focused particularly on the relevance of gamma-ray astronomy to the cosmic-ray particles which form such an important component of the 'radiation' incident on Earth and which are still of unknown origin.

The discovery of cosmic gamma rays

When Hess discovered the cosmic radiation in his perilous balloon ascents in 1912, it was natural to ascribe the so-called cosmic rays to a type of gamma radiation; the 'Hess ultra-gamma radiation'. This was because gamma rays were the most intense form of radiation known and the cosmic rays showed themselves to be very penetrating indeed. It was left to Bothe and Kolhorster nearly twenty years later to show that the cosmic radiation consisted in fact of atomic particles, as distinct from quanta, and more modern work has identified them as mainly protons with some heavier nuclei and a sprinkling of electrons and positrons. After Bothe and Kolhorster's discovery interest in cosmic gamma rays waned, although this was not due to a lack of intrinsic interest but because of the difficulty of detecting what came to be realized as a very tiny component in the primary 'beam'. The main difficulty was – and still is – that it is necessary to be very close to the top of the atmosphere (or preferably above it altogether) in order to identify the nature of the primary entity.

The importance of gamma-ray studies for the cosmic-ray physicist – if only the gamma rays can in fact, be identified – is that the rays travel in straight lines and their point of origin in the Galaxy or beyond has a good chance of being identified. This is unlike the situation for the much more abundant cosmic-ray proton component because the protons, being charged particles, are deflected by the complex and poorly known magnetic fields in the Galaxy and for the bulk of the particles, at least, their directions as determined at the Earth bear virtually no relation to the directions of their sources. Most important, from the cosmic-ray standpoint, is that many of the gamma rays are thought to be produced by cosmic rays striking gas nuclei in the interstellar medium, and the gamma rays will tell us how the cosmic-ray particles are distributed in the Galaxy, and this is vital information in any attempt to identify the origin of the particles.

Although there had been suggestions of the detection of truly cosmic gamma rays (as distinct from gamma rays generated in the atmosphere) from balloon experiments, firm detection had to await the operation of artificial satellites, and this came in March 1967 with the launch of OSO 3. Good measurements were made on gamma rays with energy about 100MeV (200 times the rest energy of the electron) and these showed a broad peak towards the galactic centre indicating that the quanta were of definite cosmic origin and that most were presumably generated in our Galaxy. The new field of gamma-ray astronomy had therefore begun.

Ways of producing cosmic gamma rays

Electromagnetic radiation is intimately connected with electrostatic charges, and clearly gamma rays will be associated with atomic nuclei and charged particles in general. At the lowest energies, in the region of MeV, radioactive nuclei are probably the most important source of gamma rays, and the resulting gamma-ray lines can give a great deal of information about the processes whereby such unstable nuclei are generated, whether they be in the outer layers of stars, supernova remnants or whatever. This subject of Gamma-Ray Line Astronomy is an even newer branch of the subject, and only during the last year have unambiguous measurements been made on these characteristic lines. This is the subject to watch in the future, but just at present there is too little information to warrant much more discussion.

Continuing to higher energies, it is likely that the gamma rays are mostly associated with cosmic-ray electrons and protons interacting either close to their sources or in the general interstellar medium. The 'target' depends on the site, but it will be either gas nuclei or radiation (or, occasionally very strong magnetic fields). Electrons have such small masses that they are easily deflected in collisions with atomic nuclei and, in so doing, if energetic enough they will emit gamma rays. Protons, on the other hand, interact by way of their short-range forces with nuclei and generate pions (the quanta of the 'nuclear glue') and some of these decay to form gamma rays.

The radiation target comprises mainly starlight and the ubiquitous 2.7 degrees Kelvin radiation that is thought to be the remains of the 'big bang' which initiated the universe. In this process – the so-called Inverse Compton interaction – energetic electrons collide with the low energy quanta of the starlight or 2.7K radiation, and increase their energy into the gamma-ray region.

The status of Gamma-ray Astronomy at present

The interesting results from OSO 3 led to bigger and better projects, the first of which was the NASA satellite SAS 2, launched on 15th November 1972. Unfortunately, a power supply fault caused it to fail in June 1973, but not before some good data had been accumulated. More recently a combined European group has launched the COS B satellite and this is still in orbit. Although COS B is providing excellent data its analysis is not as far on as that of SAS II, which is now complete, and the results from the latter form the basis for most of our discussion here.

The SAS II results are shown in Figure 1. This gives the intensity of gamma rays above 100 MeV (an energy chosen because at smaller values the accuracy of direction is poor and at higher values the number of quanta detected is low). The intensities are given as a function of galactic longitude and latitude (see Figure 2) and we see, again, as with OSO 3 that there is a broad peak in the general direction of the galactic centre. However, the shape of the longitude distribution – i.e. the way the brightness varies as we move round the galactic plane – is very different from that in the other wavelength bands. The first difference is the apparent smoothness of the distribution; with the exception of a small number of peaks the intensity varies only slowly. However, this may well be illusory in part, at least, because the accuracy of determining the gamma-ray directions is roughly $\pm 1.5°$ and very fine structure will be smoothed out. (The sky seen visually through a translucent screen giving similar smearing would obviously look very different from what we see normally and individual stars would rarely be recognized as such.) The second difference is in the large-scale structure. For example, the Sun, which is so bright optically, is nowhere to be seen in gamma rays. Furthermore, if gamma rays were distributed like starlight (but, unlike starlight

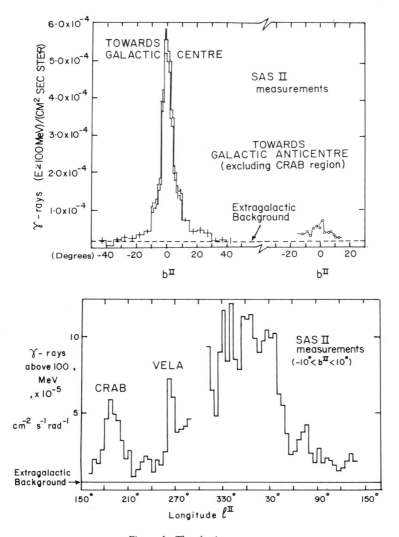

Figure 1. The sky in gamma-rays

The measurements refer to gamma-rays above 100 MeV taken with the SAS II satellite detector of Fichtel and colleagues. ℓ^{II} is the galactic longitude and b^{II} is the galactic latitude (see Fig. 2).

205

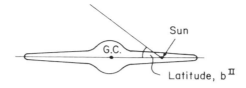

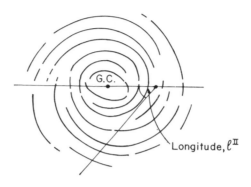

Figure 2. Galactic longitude and latitude
Schematic view of the Galaxy face-on and edge-on showing longitude and latitude.

which is absorbed by dust in the Galaxy, were able to penetrate all the interstellar medium) then there would be a much bigger peak at the galactic centre itself. Thus, as we expected from the other 'new astronomies', the view is different again through the gamma-ray window.

Looking again at the longitude distribution and the corresponding variation of intensity with latitude we see two main components, one confined largely to the Galactic plane (mainly to b II $< \pm 25°$) and identified as being of Galactic origin, and the other, which appears to be independent of latitude, and which it is tempting to class as being extragalactic in origin. Confining attention to the galactic part, there appears to be a general continuum upon which a few discrete sources are superimposed, and this is a classification which has been put forward by a number of authors who identify the continuum with gamma rays produced by cosmic-ray electrons and protons interacting with the gas nuclei in the interstellar medium, as has been remarked earlier.

Of the discrete sources two are without doubt coincident with pulsars: the Crab and Vela. Measurements by both SAS II and COS B agree that the gamma rays are pulsed at the same frequency as the radio waves in the two cases. This shows that pulsars are generating electromagnetic radiation over a very wide frequency range indeed (Figure 3) and this has added yet more urgency to the search for a complete understanding of these objects.

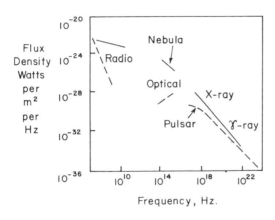

Figure 3. Electromagnetic radiation from the Crab

The COS B instrument has provided evidence for some 21 discrete sources altogether and no doubt others will be seen too. The problem concerns their identification with objects known to shine in other parts of the electromagnetic spectrum. Two more pulsars have been claimed, and one quasar (3C273), and there are suggestions that some of the others may be due to sources in HII regions – the seats of activity involved in star formation. Some of the remainder may be due to 'hot spots' in the interstellar medium where big clouds of gas molecules (Plate I) form targets for cosmic rays and thus only appear to be 'point sources'. This explanation is clearly very different from one involving individual 'small' energetic sources. Further progress in the field of identification must await long periods of operation of the COS B experiment and new satellite results.

In the meantime, we turn to an examination of the continuum which seems to form the bulk of the radiation (although we must always bear in mind the possibility that most is coming from unresolved faint discrete sources). Some support for the idea that the continuum comes mainly from cosmic-ray interactions arises if we analyse those gamma rays coming from rather close to the Earth by galactic standards (latitudes above about 10°), where we think we know the energy spectra of the initiating cosmic rays. In this region it seems that the absolute values of the measured gamma-ray intensities can be reproduced in terms of cosmic-ray interactions in the interstellar gas.

If this argument is right, then examination of the continuum gamma rays holds out the promise of shedding light on a 60-year-old problem: Where do cosmic rays come from?

Origin of the Cosmic Rays

The problem of the origin of cosmic-ray particles is a fascinating one, and one that is important for a variety of reasons, not least the question of the energetics of the universe. The energy density of cosmic rays locally (i.e. just above the earth's atmosphere) is about one electron volt per cubic centimetre, and this turns out to be very similar to that of starlight, of the magnetic field in the Galaxy, and of the average kinetic energy of gas clouds in the interstellar medium.

Thus, cosmic rays contribute considerably to the ambient energy density. If cosmic rays are produced in the Galaxy, perhaps by pulsars or supernovæ, and are trapped for long periods by the galactic magnetic field, then their number between galaxies will be much smaller than within galaxies, so that over the Universe as a whole their total energy content will presumably be similar to that contained in magnetic fields etc. If, however, cosmic rays are extragalactic in origin, and fill the whole of space, then the energy situation is very different. Now,

Plate I. Gas and dust near the young cluster M16.
Gas clouds of this type act as targets for cosmic ray protons and electrons which interact and give rise to gamma-rays.

209

their total energy content will exceed by far the other forms and only the rest energy of matter itself will be greater. The idea of an extragalactic origin is not far-fetched, and it has many adherents, who point out the fact that the intensity is nearly the same in all directions, the near equality of the energy density of the 2.7 K black-body radiation (≈ 0.25 eV per cm^3), which is certainly extragalactic, and so on.

It is here that the continuum gamma radiation holds out the promise of solving the problem at least to the extent of distinguishing between a galactic and an extragalactic origin. If the particles were extragalactic in origin, then we would expect an equal proton intensity everywhere in the Galaxy, in the absence of peculiar Galactic magnetic field configurations which would keep them out of some parts. If they were of

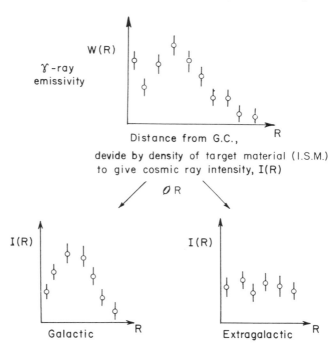

Figure 4. Principle of the method of studying cosmic rays using gamma-ray data.

galactic origin, however, then it would be expected that there would be more particles in the inner Galaxy where the density of likely sources is highest. Only in the unlikely case of the particles living so long and diffusing very far from their points of origin would we expect Galactic origin to give a uniform distribution too. If now we know how the target gas in the interstellar medium is distributed, then a knowledge of the way in which the gamma rays seen at the Earth vary in intensity with longitude and latitude is enough to allow us to derive the dependence of particle intensity on distance from the galactic centre, $I_{CR}(R)$. At this stage it should be pointed that for gamma rays above 100 MeV the cosmic rays responsible have rather low energy – the protons have energy mainly between about 1 and 10 GeV and the electrons between 0.2 and 1 GeV (1 GeV = 10^9 electron Volts).

The procedure for determining $I_{CR}(R)$ from the gamma-ray data is thus to use the observed longitude distribution of $I_\gamma(\ell)$ to derive the volume emissivity $W(R)$ (rate of production per unit volume) in the Galaxy and then to divide this by the density of target gas $8(R)$ to give $I_{CR}(R)$ (as shown in Figure 4).

The gas between the stars

Since the 1950s, measurements of the 21 cm 'song of hydrogen' have been used to map the distribution of atomic hydrogen in the Galaxy and there is a general agreement as to its form. The average density $\sigma(R)$ is shown in Figure 5. The form of $W(R)$ has been derived from the SAS II results by my colleagues A. W. Strong and D. M. Worrall, and this has been adopted. Clearly, if neutral hydrogen were the only constituent then the ensuing form of $I_{CR}(R)$ $(=W(R)/\sigma(R))$ will be a rapidly varying function of R and a Galactic origin for the cosmic-ray particles would be a certainty. Such was the situation a few years ago.

It has been known for a long time that atomic hydrogen is not the only constituent of the interstellar medium: helium and heavier nuclei have been identified, and are known to contribute another 40 per cent or so to the mass of gas. The presence of

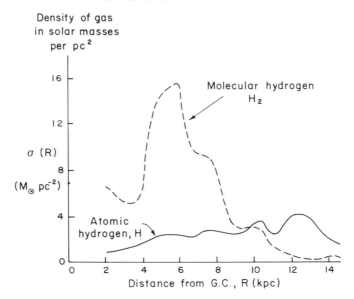

Figure 5. The density of atomic and molecular hydrogen in the Galaxy as a function of distance from the centre of the Galaxy.

The density plotted is actually the 'surface density' found by integrating over depth through the thickness of the galactic disc. It is assumed that the gas is distributed in a symmetrical way about an axis through the Galactic Centre perpendicular to the Galactic plane. The data are from the work of Gordon and Burton. (The units of density are solar masses per square parsec.)

dust is well known, too, and is the main reason for our not being able to make optical measurements on many distant stars, although its contribution to the total mass of material in the ISM is small. What was not realized until about 1975 was the extent to which hydrogen in molecular form, was a major contributor. Of course, hydrogen atoms normally form molecules very rapidly, but in the ISM, where the intensity of ultraviolet radiation is so high, the majority of H_2 molecules are quickly dissociated. Some H_2 had been detected in the ISM but its density was, quite understandably, negligible from the point of view of providing target material for cosmic-ray interactions. The surprise came from studies of a completely different gas, carbon monoxide, by way of measurements of the 2.64 mm carbon monoxide line. It is inferred that the CO molecule is excited by collisions with the unseen H_2 molecules in dense clouds of gas. The measurements made with the 36-ft radio telescope on Kitt Peak, Arizona, when transformed from CO intensities to H_2 densities, indicate that there is at least as much hydrogen in molecular form, in the Galaxy as a whole, as in its atomic form, (assuming that the conversions *are correct*). Most important for the present arguments is the way in which the molecular hydrogen is distributed in the Galaxy – as can be seen in Figure 5, there is far more H_2 than H in the important region at R~6 kpc where W (R) is greatest.

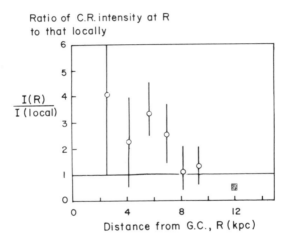

Figure 6. Distribution of Cosmic Rays in the Galaxy from the gamma ray analysis It will be noted that the uncertainties on the points are large at present (there are also the uncertainties referred to in the text arising from lack of perfect knowledge of the gas density).

Dividing W (R) by the *total* density of gas σ (R) now gives a much smaller dependence of $I_{CR}(R)$ on R (figure 6). In fact, if the H_2 densities have been underestimated by a factor 2 or so, a not impossible situation, then $I_{CR}(R)$ would be quite flat in the inner Galaxy (R<10 kpc) and our view of cosmic ray origin would have to change dramatically from galactic to extragalactic. If the densities of H_2 are correct and Figure 6 is appropriate then there is some circumstantial evidence favouring cosmic-ray production in supernovæ or pulsars but such a conclusion is not really tenable because of the H_2 problem.

A different approach is necessary.

The answer lies towards the Galactic Anticentre

If we look at Figure 5 we see that in the general direction opposite to that of the galactic centre (in the anti-centre direction, $\ell \simeq 180°$) there is much less molecular hydrogen than atomic. So, because the H_2 measurements are uncertain, we should look in this general direction if we wish to study a region where the mass of target gas is most accurately known.

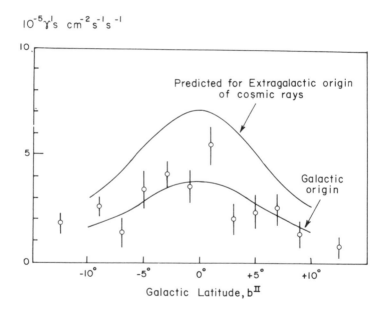

Figure 7. Gamma rays from the general direction of the Galactic Anticentre.
The data are from the SAS II experiment and refer to gamma rays above 100 MeV. The predictions are made assuming that the protons (and electrons) which interact with the ISM to produce the gamma rays are alternatively extragalactic and Galactic. It will be seen that a galactic origin is favoured.

Figure 7 shows the intensity as a function of latitude from the SAS II experiment for a rather wide range of longitudes but in the anti-centre direction in general. Also shown are curves for the rival theories of cosmic-ray particle origin – galactic and extragalactic – using the alternative cosmic-ray distributors of Figure 4 (i.e. constant for EG and varying for Galactic).

A comparison of experiment with theory shows a much better fit with the Galactic origin idea and when it is realized that if there are unidentified discrete sources also contributing then their contribution should be subtracted from the experimental intensities, the conclusion of a fall-off of I_{CR} (R) with R is strengthened further.

This then is the state of the art with Continuum Gamma Ray Astronomy at the present time – it has enabled conclusions to be made, which, although not completely watertight, do quite strongly suggest that low-energy cosmic-ray particles (i.e. particles of a few thousand MeV) are produced in our own Galaxy, This is not necessarily to suggest that all cosmic rays originate in this way and, indeed, the origin of the more energetic ones is still largely unknown. However, most of the energy of the cosmic radiation as a whole is carried by these low energy particles; and in any event, a start has been made on the 60-year-old problem.

What of the future of Gamma-Ray Astronomy?

The main problem in the subject is that of the low accuracy of measurement of gamma-ray arrival directions. An order of magnitude improvement in precision will surely throw up many new sources and their study, together with further work on the residual continuum, will be most rewarding.

Finally, there is the whole question of the apparently isotropic background radiation (Figure 1) which many workers believe to be extragalactic in character. If the present preliminary measurements are correct, then its intensity is about ten times what would have been expected on the basis of the summed contribution from all other galaxies like our own. There is thus evidence for other objects in the Universe more energetic than our own Galaxy – a situation seen at other

215

wavelengths too – and this is clearly another growth point. Already a few other galaxies have been seen in gamma rays and improved precision will, without doubt, indicate others. The situation about the extragalactic continuum (i.e. the non discrete source contribution) is very similar to that with the galactic continuum – there is evidence for it but it is not yet completely firm. If it is genuine, then a cosmological origin is likely, and already a number of theories have been put forward, the most interesting being that it derives from the annihilation of matter and antimatter generated in the big bang.

Gamma-Ray Astronomy obviously has an exciting future ahead of it.

Recent Advances in Astronomy

PATRICK MOORE

In reviewing the period between mid-1978 and mid-1979, one has to admit that developments in space research dominate the picture. In the Solar System, however, there has been one interesting Earth-based discovery – reported briefly in the 1979 *Yearbook* as a 'stop press' item. This concerns the satellite of Pluto.

PLUTO

Pluto has long been an enigma. If it is to be regarded as a bona-fide planet, its orbit is wrong, its size is wrong, and its mass is wrong. Since last January (1979) it has temporarily forfeited its title of 'the outermost planet', because it has moved within the orbit of Neptune. Whether or not the successful prediction of its position by Lowell – and, independently, by Pickering – can have been pure luck is still a matter for debate.

Photographs taken with the large reflector at the U.S. Naval Observatory showed that there was something curious about Pluto's appearance, and there now seems little doubt that we are dealing with a kind of binary system. The secondary body – unofficially named Charon, after the boatman of the Styx – is too large, compared with Pluto itself, to be classed as a mere satellite. But it now seems that even Pluto is much smaller and less massive than previously believed; the diameter may be no more than 1800 miles, less than that of our Moon, while Charon's diameter is 500 or 600 miles at least. Moreover, the combined masses of the two bodies are very low, and their densities are comparable with that of water. Pluto and Charon have, in fact, been likened to two large, lonely ice-balls.

Obviously these results are preliminary only, and it is unfortunate that no Pluto probe has been planned as yet; but it does look as though conditions in the far reaches of the Solar System are peculiar by any standards, and more and more astronomers are coming to the conclusion that a more remote planet may exist, though it will certainly be faint and extremely hard to locate.

VENUS

Much nearer home, we have had a positive fleet of Venus probes, all of which reached their target in December 1978. Two were American – one of them a 'multiprobe', consisting of a 'bus' and several smaller vehicles – and two were Russian landers. The results show that Venus is every bit as strange as had been expected. A huge rift valley has been identified, 900 miles long, 175 miles wide in places and about three miles deep; there is a virtually cloudless 'hole' in the atmosphere over the planet's north pole; and anyone on the surface of Venus would be subjected to continuous lightning flashes, accompanied by thunder. Only about half an hour after arrival, the Soviet Venera 11 used its acoustic equipment to detect a loud noise reaching 82 db, which was presumably thunder. The temperature at the surface was given as 854° F, with a ground pressure of 90.5 atmospheres. I fear that we must, for the moment at least, abandon any idea of sending astronauts to Venus!

JUPITER

Jupiter has also been very much in the news, thanks to the two Voyager probes, the first of which made its pass of the planet in March and the other in July. Exquisite pictures of Jupiter itself were obtained, notably of the stormlike, spinning Red Spot, and results from the night side revealed lightning flashes together with a long auroral arch. Jupiter, too, has its thunderstorms, and may be classed as a noisy world as well as a lethal one. A narrow, obscure ring was found, but the most striking news concerned the satellites – four of which, remember, are large enough to be seen with virtually any optical equipment (I know a few lynx-eyed people who can glimpse them with the naked eye). Callisto and Ganymede are the low-density, 'icy' satellites, with craters aplenty; Callisto is the most heavily-cratered object so far found. Europa is denser, but it is Io which really catches the imagination. Its red, sulphury surface has been compared with a pizza; there are no craters, but there are active volcanoes, some of which were photographed in eruption. The cause of this violent vulcanism has caused great discussion, but no doubt Jupiter's tidal pull, plus radiation bombardment, has much to do with it. Amalthea, the tiny innermost satellite, has also been studied from close range; it is red, and irregular in shape.

SATURN

At the time of writing (July 1979) both the Voyagers are en route for Saturn, and will arrive near that planet in 1980 and 1981 respectively. We also have Pioneer 11, which made its pass of Jupiter as long ago as December 1974, and is due to rendezvous with Saturn on 1 September 1979, passing some distance outside the main ring system. By the time that this *Yearbook* appears in print, we will know how successful (or otherwise) it has been, and I assure you that any interesting results will be reported in our next issue!

SKYLAB

Artificial satellites of many kinds continue to be sent up, but one of them – the U.S. space-station Skylab – came to an inglorious end in July 1979, when it plunged back into the Earth's atmosphere and disintegrated. This happened rather before the NASA authorities had expected, though it must be added that Desmond King-Hele, the British satellite tracker, had his sums right. The accelerated delay was due to solar activity. When the Sun becomes really energetic, there are effects on the Earth's upper air, which thickens somewhat. This duly happened as the Sun worked up to its expected 1979-80 maximum, and the drag on Skylab was increased sufficiently to make it re-enter. Had it lasted until 1983, as the NASA authorities had originally expected, there would have been hopes of sending a Shuttle to it and boosting it into a higher, permanent orbit. Many people will regret its demise; it has an honoured place in history – and if it could have been saved, it would have made an excellent space museum for our descendants.

The Shuttle itself is in an advanced stage of planning, and may have been tested before the end of 1979. It is, of course, a recoverable vehicle, and should make space-travel to and from orbital stations not only cheaper, but a great deal safer. Meantime, one unmanned satellite of special interest is Britain's Aeriel 6, which was put safely into its orbit in midsummer with the set aim of improving our knowledge of cosmic-ray and X-ray astronomy. X-rays from space cannot be studied from ground level, owing to the effective if annoying screen provided by our atmosphere; Ariel 5, the previous British satellite, had made great progress, and Ariel 6 was designed to capitalize upon what had been learned. For instance, it is thought that X-rays come from the centres of those huge assemblages of stars known as globular clusters, and there have been suggestions that the root cause of such a phenomenon is a Black Hole in the cluster centre – greedily gobbling up material and giving rise to X-radiation in the process, since the material is raised to a high temperature before being consumed. Whether or not this idea is valid remains to be seen, but it is an intriguing possibility.

ISAAC NEWTON TELESCOPE MOVE

Finally, it is worth mentioning a notable telescope transfer.

The largest telescope ever set up in the British Isles has been the 98-inch Isaac Newton reflector at the Royal Greenwich Observatory, Herstmonceux. Of course it is very gratifying to have such a giant telescope near at hand, but nobody could pretend that the climate is ideal, and so the I.N.T was dismantled in the summer of 1979 for its move to the new Northern Hemisphere Observatory being established at La Palma, in the Canary Islands (not Las Palmas, please note). The I.N.T. will be only second in size of the Observatory telescopes, but all in all most people consider that scientifically the decision was a wise one. As an observing site, Sussex can hardly be classed with California, Chile, Australia, or South Africa. Rather sadly, therefore, the staff of Greenwich waved good-bye to their cherished I.N.T., and wished it all success in its new home.

PART THREE

Miscellaneous

Some Interesting Telescopic Variable Stars

Star	R.A.		Dec.		Mag. range	Period, days	Remarks
	h	m	°				
R. Andromedæ	0	22	+38	18	6.1–14.9	409	
W Andromedæ	2	14	+44	4	6.7–14.5	397	
Theta Apodis	14	00	−76	33	6.4– 8.6	119	Semi-regular.
R Aquilæ	19	4	+ 8	9	5.7–12.0	300	
R Arietis	2	13	+24	50	7.5–13.7	189	
R Aræ	16	35	−56	54	5.9–6.9	4	Algol type.
R Aurigæ	5	13	+53	32	6.7 13.7	459	
R Boötis	14	35	+26	57	6.7 12.8	223	
Eta Carinae	10	43	−59	25	−0.8– 7.9	—	Unique erratic variable.
l Carinæ	09	43	−62	34	3.9 10.0	381	
R Cassiopeiæ	23	56	+51	6	5.5 13.0	431	
T Cassiopeiæ	0	20	+55	31	7.3–12.4	445	
X Centauri	11	46	−41	28	7.0–13.9	315	
T Centauri	13	38	−33	21	5.5– 9.0	91	Semi-regular.
T Cephei	21	9	+68	17	5.4 11.0	390	
R Crucis	12	20	−61	21	6.9– 8.0	5	Cepheid.
Omicron Ceti	2	17	− 3	12	2.0–10.1	331	Mira.
R Coronæ Borealis	15	46	+28	18	5.8–14.8	–	Irregular
W Coronæ Borealis	16	16	+37	55	7.8–14.3	238	
R Cygni	19	35	+50	5	6.5–14.2	426	
U Cygni	20	18	+47	44	6.7–11.4	465	
W Cygni	21	34	+45	9	5.0– 7.6	131	
SS Cygni	21	41	+43	21	8.2–12.1	–	Irregular.
Chi Cygni	19	49	+32	47	3.3–14.2	407	Near Eta.
Beta Doradûs	05	33	−62	31	4.5– 5.7	9	Cepheid.
R Draconis	16	32	+66	52	6.9–13.0	246	
R Geminorum	7	4	+22	47	6.0–14.0	370	
U Geminorum	7	52	+22	8	8.8–14.4		Irregular.
R Gruis	21	45	−47	09	7.4 14.9	333	
S Gruis	22	23	−48	41	6.0 15.0	410	
S Herculis	16	50	+15	2	7.0 13.8	307	
U Herculis	16	23	+19	0	7.0 13.4	406	
R Hydræ	13	27	−23	1	4.0–10.0	386	
R Leonis	9	45	+11	40	5.4–10.5	313	Near 18, 19.
X Leonis	9	48	+12	7	12.0–15.1	–	Irregular (U Gem type).
R Leporis	4	57	−14	53	5.9–10.5	432	'Crimson star.'
R Lyncis	6	57	+55	24	7.2–14.0	379	
W Lyræ	18	13	+36	39	7.9–13.0	196	
T Normæ	15	40`	−54	50	6.2–13.4	293	
HR Delphini	20	40	+18	58	3.6– ?	–	Nova, 1967.
S Octantis	17	46	−85	48	7.4–14.0	259	
U Orionis	5	53	+20	10	5.3–12.6	372	
Kappa Pavonis	18	51	−67	18	4.0– 5.5	9	Cepheid.
R Pegasi	23	4	+10	16	7.1–13.8	378	
S Persei	2	19	+58	22	7.9–11.1	810	Semi-regular.
R Sculptoris	01	24	−32	48	5.8– 7.7	363	Semi-regular.

Star	R.A.		Dec.		Mag. range	Period days	Remarks
	h	m	°	'			
R Phœnicis	23	53	−50	05	7.5–14.4	268	
Zeta Phœnicis	01	06	−55	31	3.6– 4.1	1	Algol type.
R Pictoris	04	44	−49	20	6.7–10.0	171	Semi-regular.
L² Puppis	07	12	−44	33	2.6– 6.0	141	Semi-regular.
Z Puppis	07·	30	−20	33	7.2–14.6	510	
T Pyxidis	09	02	−32	11	7.0–14.0	–	Recurrent nova (1920, 1944)
R Scuti	18	45	− 5	46	5.0– 8.4	144	
R Serpentis	15	48	+15	17	5.7–14.4	357	
SU Tauri	5	46	+19	3	9.2–16.0	–	Irregular (R CrB type).
R Ursæ Majoris	10	41	+69	2	6.7–13.4	302	
S Ursæ Majoris	12	42	+61	22	7.4–12.3	226	
T Ursæ Majoris	12	34	+59	46	6.6–13.4	257	
S Virginis	13	30	-6	56	6.3–13.2	380	
R Vulpeculæ	21	2	+23	38	8.1–12.6	137	

Note: Unless otherwise stated, all these variables are of the Mira type.

Some Interesting Double Stars

The pairs listed below are well-known objects, and all the primaries are easily visible with the naked eye, so that right ascensions and declinations are not given. Most can be seen with a 3-inch refractor, and all with a 4-inch under good conditions, while quite a number can be separated with smaller telescopes, and a few (such as Alpha Capricorni) with the naked eye. Yet other pairs, such as Mizar-Alcor in Ursa Major and Theta Tauri in the Hyades, are regarded as too wide to be regarded as bona-fide doubles!

Name	Magnitudes	Separation"	Position angle, deg.	Remarks
Gamma Andromedæ	3.0, 5.0	9.8	060	Yellow, blue. B is again double (0".4) but needs a larger telescope.
Zeta Aquarii	4.4, 4.6	2.6	291	Becoming more difficult.
Gamma Arietis	4.2, 4.4	8	000	Very easy.
Theta Aurigæ	2.7, 7.2	3	330	Stiff test for 3 in. OG
Delta Boötis	3.2, 7.4	105	079	Fixed.
Epsilon Boötis	3.0, 6.3	2.8	340	Yellow, blue. Fine pair.
Kappa Boötis	5.1, 7.2	13	237	Easy.
Zeta Cancri	5.6, 6.1	5.6	082	
Iota Cancri	4.4, 6.5	31	307	Easy. Yellow, blue.
Alpha Canum Venat.	3.2, 5.7	20	228	Yellowish, bluish. Easy.
Alpha Capricorni	3.3, 4.2	376	291	Naked-eye pair. Alpha again double.
Eta Cassiopeiæ	3.7, 7.4	11	298	Creamy, bluish. Easy.
Beta Cephei	3.3, 8.0	14	250	
Delta Cephei	var, 7.5	41	192	Very easy.

Name	Magnitudes	Separation"	Position angle, deg.	Remarks
Alpha Centauri	0.0, 1.7			Binary; period 80 years. Very easy.
Xi Cephei	4.7, 6.5	6	270	Reasonably easy.
Gamma Ceti	3.7, 6.2	3	300	Not too easy.
Alpha Circini	3.4, 8.8	15.8	235	PA, slowly decreasing.
Zeta Coronæ Borealis	4.0, 4.9	6.3	304	
Delta Corvi	3.0, 8.5	24	212	
Alpha Crucis	1.6, 2.1	4.7	114	Third star in low-power field.
Gamma Crucis	1.6, 6.7	111	212	Wide optical pair.
Beta Cygni	3.0, 5.3	35	055	Yellow, green. Glorious.
61 Cygni	5.3, 5.9	25	150	
Gamma Delphini	4.0, 5.0	10	265	Yellow, greenish. Easy.
Nu Draconis	4.6, 4.6	62	312	Naked-eye pair.
Alpha Geminorum	2.0, 2.8	2	151	Castor. Becoming easier.
Delta Geminorum	3.2, 8.2	6.5	120	
Alpha Herculis	var, 6.1	4.5	110	Red, green.
Delta Herculis	3.0, 7.5	11	208	Optical double.
Zeta Herculis	3.0, 6.5	1.4	300	Fine, rapid binary.
Gamma Leonis	2.6, 3.8	4.3	121	Binary; period 400 years
Alpha Lyræ	0.0, 10.5	60	180	Vega. Optical; B faint.
Epsilon Lyræ	4.6, 6.3	3	005	Quadruple. Both pairs
	4.9, 5.2	2.3	111	separable in 3 in. OG
Zeta Lyræ	4.2, 5.5	44	150	Fixed. Easy double.
Beta Orionis	0.1, 6.7	9.5	205	Rigel. Can be split with 3 in.
Iota Orionis	3.2, 7.3	11	140	
Theta Orionis	6.0, 7.0			The famous Trapezium in M.42
	7.5, 8.0			
Sigma Orionis	4.0, 7.0	11.1	236	Quadruple. D is rather
		12.9	085	faint in small apertures.
Zeta Orionis	1.9, 5.0	3	160	
Eta Persei	4.0, 8.5	28.5	300	Yellow, bluish.
Beta Phoenicis	4.1, 4.1	1.3	352	Slow binary.
Beta Piscis Austrini	4.4, 7.9	30.4	172	Optical pair. Fixed.
Alpha Piscium	4.3, 5.3	1.9	291	
Kappa Puppis	4.5, 4.6	9.8	318	Again double.
Alpha Scorpii	0.9, 6.8	3	275	Antares. Red, green.
Nu Scorpii	4.2, 6.5	42	336	
Theta Serpentis	4.1, 4.1	23	103	Very easy.
Alpha Tauri	0.8, 11.2	130	032	Aldebaran. Wide, but B is very faint in small telescopes.
Beta Tucanæ	4.5, 4.5	27,1	170	Both components again double.
Zeta Ursæ Majoris	2.3, 4.2	14.5	150	Mizar. Very easy. Naked eye pair with Alcor.
Alpha Ursæ Minoris	2.0, 9.0	18.3	217	Polaris. Can be seen with 3 in.
Gamma Virginis	3.6, 3.7	4.8	305	Binary; period 180 yrs. Closing.
Theta Virginis	4.0, 9.0	7	340	Not too easy.
Gamma Volantis	3.9, 5.8	13.8	299	Very slow binary.

Some Interesting Nebulae and Clusters

Object	R.A.		Dec.		Remarks
	h	m	°		
M.31 Andromedæ	00	40.7	+41	05	Great Galaxy, visible to naked eye.
H.VIII 78 Cassiopeiæ	00	41.3	+61	36	Fine cluster, between Gamma and Kappa Cassiopeiæ.
M.33 Trianguli	01	31.8	+30	28	Spiral. Difficult with small apertures.
H.VI 33-4 Persei	02	18.3	+56	59	Double cluster; Sword-handle.
△142 Doradûs	05	39.1	−69	09	Looped nebula round 30 Doradûs. Naked-eye. In Large Cloud of Magellan.
M.1 Tauri	05	32.3	+22	00	Crab Nebula, near Zeta Tauri.
M.42 Orionis	05	33.4	−05	24	Great Nebula. Contains the famous Trapezium, Theta Orionis.
M.35 Geminorum	06	06.5	+24	21	Open cluster near Eta Geminorum.
H.VII 2 Monocerotis	06	30.7	+04	53	Open cluster, just visible to naked eye.
M.41 Canis Majoris	06	45.5	−20	42	Open cluster, just visible to naked eye.
M.47 Puppis	07	34.3	−14	22	Mag. 5,2. Loose cluster.
H.IV 64 Puppis	07	39.6	−18	05	Bright planetary in rich neighbourhood.
M.46 Puppis	07	39.5	−14	42	Open cluster.
M.44 Cancri	08	38	+20	07	Præsepe. Open cluster near Delta Cancri. Visible to naked eye.
M.97 Ursæ Majoris	11	12.6	+55	13	Owl Nebula, diameter 3'. Planetary.
Kappa Crucis	12	50.7	−60	05	"Jewel Box"; open cluster, with stars of contrasting colours.
M.3 Can. Ven.	13	40.6	+28	34	Bright globular.
Omega Centauri	13	23.7	−47	03	Finest of all globulars. Easy with naked eye.
M.80 Scorpii	16	14.9	−22	53	Globular, between Antares and Beta Scorpionis.
M.4 Scorpii	16	21.5	−26	26	Open cluster close to Antares.
M.13 Herculis	16	40	+36	31	Globular. Just visible to naked eye.
M.92 Herculis	17	16.1	+43	11	Globular. Between Iota and Eta Herculis.
M.6 Scorpii	17	36.8	−32	11	Open cluster; naked-eye.
M.7 Scorpii	17	50.6	−34	48	Very bright open cluster; naked eye.
M.23 Sagittarii	17	54.8	−19	01	Open cluster nearly 50' in diameter.
H.IV 37 Draconis·	17	58.6	+66	38	Bright Planetary.
M.8 Sagittarii	18	01.4	−24	23	Lagoon Nebula. Gaseous. Just visible with naked eye.
NGC 6572 Ophiuchi	18	10.9	+06	50	Bright planetary, between Beta Ophiuchi and Zeta Aquilæ.
M.17 Sagittarii	18	18.8	−16	12	Omega Nebula. Gaseous. Large and bright.
M.11 Scuti	18	49.0	−06	19	Wild Duck. Bright open cluster.
M.57 Lyræ	18	52.6	+32	59	Ring Nebula. Brightest of planetaries.
M.27 Vulpeculæ	19	58.1	+22	37	Dumb-bell Nebula, near Gamma Sagittæ.
H.IV 1 Aquarii	21	02.1	−11	31	Bright planetary near Nu Aquarii.
M.15 Pegasi	21	28.3	+12	01	Bright globular, near Epsilon Pegasi.
M.39 Cygni	21	31.0	+48	17	Open cluster between Deneb and Alpha Lacertæ. Well seen with low powers.

Some Recent Books

The Crab Nebula, by Simon Mitton. Faber & Faber, London and Boston 1979. A detailed Study of this remarkable object, intelligible to the beginner and of value to the expert.

In The Centre of Immensities, by Bernard Lovell, Hutchinson, London 1979. A thought-provoking book, dealing with cosmology, the extent of life in the Universe, and the future of mankind.

Man and the Stars, by Hanbury Brown. The effects on human thought due to the stars – the calendar, time, navigation, science, and Man himself.

The Road to the Stars, by Iain Nicolson. David & Charles, Newton Abbot and Vancouver, 1978. A fascinating analysis of the possibilities of interstellar travel.

The Guinness Book of Astronomy Facts and Feats, by Patrick Moore. Guinness Superlatives, London 1979. Data, history, details of the bodies in the Solar System, star maps, and lists, etc.

Our Contributors

IOSIF SHKLOVSKY, Corresponding Member of the U.S.S.R. Academy of Sciences, is one of the leading Soviet astronomers. His outstanding work upon subjects such as the synchrotron radiation from the Crab Nebula has been universally recognized. He is co-author with Carl Sagan of a well-known book, *Intelligent Life in the Universe.*

D. L. McNAUGHTON carries out his researches at the Meteorological Department in Salisbury, Rhodesia. He has paid careful attention to the possible association between meteors and rainfall.

PETER J. GARBETT is a young member of the British Astronomical Association He has his private observatory at Barton-le-Clay in Bedfordshire, and is particularly interested in studies of the Sun and the Moon.

DR GARRY E. HUNT formerly of the Meteorological Office (Bracknell) and now of University College, London, is one of the world's leading experts in planetary research, and is closely concerned with current space-probe projects, so that he makes frequent visits to American centres. In addition to his many technical contributions he is also well known for his television broadcasts and popular lectures.

PROFESSOR ÅKE WALLENQUIST is a leading Swedish astronomer, now Professor Emeritus at the Observatory of Uppsala. He has published popular books as well as important research papers, and is a well-known broadcaster in Sweden. He has paid special attention to problems of star-clusters.

MARTIN COHEN is a Cambridge graduate who has been carrying out researches in the United States. He has concentrated upon infrared astronomy, to which subject he has made important contributions.

DAVID A. ALLEN (one of our most regular and welcome contributors!) took his B.A. at Cambridge University in 1967, and his PhD. in 1971. After some years at the Royal Greenwich Observatory, he is now continuing his research in Australia. His recent work has concentrated upon infrared astronomy, for which he has used some of the world's most powerful telescopes, including the Palomar 200-inch.

PROFESSOR A.W. WOLFENDALE, F.R.S., is Professor of Physics at the University of Durham. His research work extends over many branches of physics and astronomy, including cosmic radiation and gamma rays.